Artificial Reproduction:
A Social Investigation

Artificial Reproduction:
A Social Investigation

R. SNOWDEN, G. D. MITCHELL and
E. M. SNOWDEN

Institute of Population Studies, University of Exeter

London
GEORGE ALLEN & UNWIN
Boston   Sydney

George Allen & Unwin (Publishers) Ltd,
40 Museum Street, London WC1A 1LU, UK

George Allen & Unwin (Publishers) Ltd,
Park Lane, Hemel Hempstead, Herts HP2 4TE, UK

Allen & Unwin, Inc.,
9 Winchester Terrace, Winchester, Mass. 01890, USA

George Allen & Unwin Australia Pty Ltd,
8 Napier Street, North Sydney, NSW 2060, Australia

First published in 1983

British Library Cataloguing in Publication Data

Snowden, Robert
    Artificial reproduction.
1. Fertilization in vitro, Human
I. Title   II. Mitchell, G. Duncan
III. Snowden, E.M.
618.1'78059   RG135
ISBN 0-04-176003-4

Library of Congress Cataloging in Publication Data

Snowden, R. (Robert)
    Artificial reproduction.
Bibliography: p.
Includes index.
1. Artificial insemination, Human–Social aspects. I. Mitchell, G. Duncan
(Geoffrey Duncan). II. Snowden, E. M. III. Title. [DNLM: 1. Insemina-
tion, Artificial. 2. Sociology. HQ 761 S674e]
RG134.S64 1983   618.2   83-11809
ISBN 0-04-176003-4

Set in 11 on 13 point Garamond by Bedford Typesetters Ltd
and printed in Great Britain
by Biddles Ltd, Guildford, Surrey

Contents

Contents

Preface

For the first time in the history of mankind it is now possible
to conceive and begin the development of children without
the need for sexual intercourse. The implications of this bold
statement are far-reaching and poorly understood. There is
no human precedent for such an event and the pace of
technological advance in reproductive biology is so rapid
that the future can only vaguely be discerned, for even as we
write further developments are taking place which introduce
completely new issues. Whilst in recent decades we have had
to come to terms with technological advances which have
brought about significant social changes, for example the jet
engine, nuclear fission and the micro-chip, none of them
affect the very basis of social organisation as does the freeze-
storing of ova, sperm and embryos, and AID, *'in-vitro'*
fertilisation and embryo transfer. We call these procedures
artificial reproduction. Some people may argue that the word
'artificial' indicates a negative or pejorative view of the
subject, but the word simply implies a creation by human
skill, and this is what we intend by the term.

Every technological development has some effect upon
the relationships people have with one another and artificial
reproduction is certainly no exception. Indeed, human
relationships do not exist in isolation; they provide the
matrix upon which human society is based. The confusion
created by artificial reproduction arises from uncertainty
about the description and quality of those relationships
contained in the basic unit of social organisation – the family.
This uncertainty is aggravated by the lack of an adequate
terminology, for whilst everyday speech may take us so far

in describing these relationships, there comes a time when
we need to use a new and specific set of terms to describe the
various parts being played by the actors on this human stage;
this we have attempted to do. However, the particular
problem that has faced us is the task of doing justice to both
the individual and the society in which he or she belongs. In
this book we summarise our research findings and in the last
chapter make some recommendations. In doing so we are
well aware of the difficulty of finding a balance between the
needs and desires of individuals and the welfare of society
generally. The difficulty is not made any easier by the
prevailing cultural bias of the Western world, which
emphasises individualistic values. Recognising a plurality of
values in our society, we acknowledge that obtaining a
consensus is difficult. Thus in the world in which we live
there must of necessity be an accommodation to a variety of
often conflicting interests; individual, professional and
political. Moreover, these conflicts are to be found in sexual
behaviour and family life, including the raising of children as
well as in other spheres. Nevertheless, we do discern a
substratum of convictions which parents share about the
nurture and upbringing of children. It is with these convictions in mind, and here we refer especially to fundamental
values such as honesty, truthfulness and integrity, that we
have embarked on this study of artificial reproduction.

Our research has concentrated on the most common form
of artificial reproduction, namely, artificial insemination by
donor semen (AID), but our interests have also extended to
the social issues surrounding other forms of artificial reproduction. Indeed, we believe that an understanding of the
social issues surrounding AID will provide the key to an
understanding of these other forms. We have reviewed the
literature, considered 1,000 cases of AID, and have had
discussions in depth with over sixty couples who are parents
of one or more AID children. We have also talked with older
AID children themselves, with parents of grown-up AID

children, and with AI donors and practitioners. Whereas in our first book, *The Artificial Family: A Consideration of Artificial Insemination by Donor*, we gave our reflections on the subject, in this book we are pleased to be able to give some account of our researches and our findings. It is this that enables us to make some recommendations. We do so in the knowledge that the subject is a topical one and that practical decisions urgently need to be made.

R. SNOWDEN   G. D. MITCHELL   E. M. SNOWDEN

Institute of Population Studies, University of Exeter, February 1983

Acknowledgements

The authors wish to thank most sincerely those who have given them their time and allowed them to invade their privacy so that a greater understanding of the social implications of artificial reproduction could be obtained. They acknowledge gratefully a grant for research by the Social Science Research Council.

The study which has given rise to this book could not have been carried out without the foresight, initiative and co-operation of Dr Margaret Jackson, Hon. FRCOG, a pioneer in family planning, and the devoted work of her secretary Miss Muriel Bartlett. Our thanks are extended to them and also to the staff of the Institute of Population Studies, Exeter University, in particular to Mrs Ann McClary, and Mrs Ruth Preist.

*Artificial Reproduction:
A Social Investigation*

Part I

ARTIFICIAL REPRODUCTION AND THE FAMILY

1

The Separation of Reproduction from Sexual Intercourse

Most people would find little difficulty in recognising that the birth of a child is an occurrence of major social significance. Since the beginnings of human life when people grouped themselves together for protection, economic necessity or for any other reason, reproduction has been more than merely a biological event of concern only to the mother and child. Nor is the social importance of birth confined to a small number of closely related people; all members of a society are affected in some way by the number and health of the children being born into that society.

Apart from the excesses occasionally found where a direct policy of sterilisation of selected, and presumably unwanted, minority groups living within a more broadly based society is undertaken, the control of reproduction is not usually a direct one. Reproduction is controlled indirectly by defining who is permitted to have sexual intercourse with whom and in what circumstances. This control over reproduction is based upon the assumption that sexual intercourse leads to reproduction and that by controlling sexual intercourse

3

reproduction can be effectively channelled in directions that maintain the society as a whole. The control over sexual intercourse can be of a formal kind as is found in the laws relating to incest and to the age of consent or of an informal nature where couples are made to feel 'different' if they produce too few or too many children.

Another assumption concerns the relationship between marriage and sexual intercourse. For centuries it has been the socially accepted view that sexual intercourse should only take place within marriage. Whilst recognising that sexual intercourse does take place before and outside marriage, the sentiment still expressed by most people in our own society is that it would be preferable if these extra-marital sexual activities did not take place. The reason for this view is a traditional one and some would say that its relevance to the late twentieth century is limited especially when techno-logical advances in relation to the control of fertility have tended to separate sexual intercourse from reproduction.

The separation of sexual intercourse from reproduction was the first major change which affected the social rules governing who should have sexual intercourse with whom. Until effective contraception had arrived, the distinction between sexual intercourse and reproduction was limited. Indeed, for 'respectable' people at least, the pursuit of reproduction was what sexual intercourse was about. It is true that sexual intercourse for 'fun' did occur in earlier times but sexual intercourse within marriage generally had but one result. The arrival of contraception heralded the possibility that sexual intercourse, even within marriage, could take place independently of the actuality or fear of pregnancy. After a shaky start when most informed people were opposed to contraception, the practice is now so widespread that the experience of an unplanned pregnancy is usually regarded as resulting from a lack of adequate sex education, the poor provision of family planning services or an unwillingness to accept the advice or service offered. In other words, the relationship between sexual intercourse and reproduction had altered. Instead of sexual intercourse leading to reproduction it can now be planned as an experience in its own right. This, we believe, has inexorably altered the relationship between men and women, and husbands and wives, in ways that are only now becoming apparent. A consideration of the differences between the views of those women who led the way for the emancipation of women in general at the turn of the present century with those fighting for similar aims in the 1960s and later, makes this point very clearly. Whereas the earlier movement campaigned for less sexual demands by husbands on their wives, those who came later, armed with the means of preventing pregnancy by the use of contraceptives, were demanding equal sexual status with men. The separation of sexual intercourse as 'fun' from sexual intercourse for reproduction remains very significant. It is not that reproduction is no longer desired but that its control in terms of timing and frequency is now more possible than previously.

If sexual intercourse can now take place without the risk of reproduction, the argument that society as well as those directly involved has some interest in the matter loses much of its force. Without the risk of pregnancy and the birth of a child sexual intercourse has become a private and personal activity that should concern no-one other than the couple involved.

Yet another step has taken place which adds even further complication. Reproduction need now no longer be preceded by sexual intercourse. The techniques of what we collectively term 'artificial reproduction' permit the ovum and the sperm to meet and to begin the development of a new human being independently of the physical union of man and woman. The attendant changes in social attitudes and behaviour this new technology will doubtless bring about, can only vaguely be discerned at the present time. One issue is nevertheless clear: if reproduction was previously controlled indirectly by determining who could have sexual intercourse with whom, then the new techniques effectively destroy that control. It may be that more direct control over reproduction, at least of children produced by means other than sexual intercourse, will be required if society as we know it today is not to be significantly changed for the generations yet to be born. We will return to this again after a brief review of how the separation of reproduction from sexual intercourse has come about; a discussion of why some of the procedures used are accompanied by features which may be disruptive to family life; and a review of the experience of hundreds of couples who have used one particular form of artificial reproduction. The argument presented in this book is that some form of social control of the artificial reproduction techniques and their associated technological developments is desirable in order to protect the individuals directly involved, the institution of family life and the social structure of society in general.

Artificial reproduction techniques are being used with increasing frequency to enable childless couples to overcome their infertility. These techniques include artificial insemination by husband (AIH) and by donor (AID), surrogate motherhood and external human fertilisation (commonly called '*in-vitro*' fertilisation leading to the birth of a 'test-tube' baby). Accompanying these techniques other scientific advances have taken place including cryopreservation or freeze-storage of sperm, ova and even embryos which have considerably increased the possibilities for use of the techniques themselves.

All the forms of artificial reproduction described have but one feature in common: they effectively separate reproduction from sexual intercourse. We have indicated that this apparently simple statement raises social, moral and ethical questions which go far beyond the technological and medical considerations that have received most of the publicity to date. Herein lies the dilemma. The desire to assist the childless couple or woman has to be balanced against the implications for society as a whole. This does not mean that an 'all or nothing' situation need develop, for balance can be achieved by the institution of agreed controls and the creation of a structure within which appropriate safeguards can operate. The heartbreak of involuntary childlessness is difficult to exaggerate; techniques which enable these couples and individuals to become parents are to be welcomed and encouraged. Nevertheless these techniques, the scientific possibilities associated with their use, and their attendant behavioural implications need to be identified and understood if societal needs are also to be met.

In order to examine the social significance of the separation of reproduction from sexual intercourse, it is necessary to consider the various forms of artificial reproduction and then to identify their relative effect on the social system in which we all lead our lives. This is not a simple task, for merely listing a series of chronological events or itemising the progress of technological achievements is insufficient. As a particular technological development is achieved, its use within society may have repercussions which far exceed the complexity of the development itself. For example, the technique of clinically introducing semen into the vagina is relatively simple but its social implications may be profound depending upon the relationship between the woman receiving the semen and the man providing it in the first place. If they are related by marriage the social implications of their involvement in the procedure are of a different order compared to the situation where they are unrelated or even unknown to each other. Again, the technique of fertilisation outside the human body and the freeze-storage of the resulting embryo may be examples of greater scientific advance, but their social implications are so profound that ultimately these are more important than the technical and scientific achievement itself.

It is with these implications in mind that the alternative forms of artificial reproduction are described. Ours is not a technical, legalistic or moralistic interpretation but a social one. This is not to deny the role of each of these other important approaches, or to suggest that a social perspective can effectively be undertaken without reference to them. Our emphasis at all times is to view the practice of artificial reproduction in terms of its *social* implications.

ARTIFICIAL INSEMINATION USING THE HUSBAND'S SEMEN (AIH)

This is the oldest form of artificial reproduction in humans. Artificial insemination by husband (AIH) takes place when the husband's sperm is delivered into the body of his wife by means other than sexual intercourse. The first documented case of AIH occurred some time during the 1770s. A recently married London cloth merchant who suffered from hypospadias consulted John Hunter in the hope of overcoming this disability and fathering a child. Hunter provided a syringe which was to be warmed and filled with the semen which escaped during coitus; the semen was then to be injected into the vagina. This was done and the merchant's wife bore a child. It is not certain whether Hunter performed the AIH himself, or whether he merely advised it.

Today AIH may be indicated by abnormalities occurring in either the husband or the wife, and is an attempt to increase the chances of fertilisation. For example, fertile semen may fail to be delivered to the cervix because of an anatomical anomaly such as hypospadias, because of impotence, or because of vaginal spasm. Among such couples any simple technique that delivers the husband's sperm to the cervix of his wife is likely to prove successful. Couples can be provided with a syringe or cervical insemination cap and carry out the procedure themselves at home.

AIH may also be indicated where the semen is subfertile due to small numbers of spermatozoa, or to sperm which are

8

malformed or relatively inactive. In some cases the wife's cervical mucus may be of a consistency which is impenetrable or hostile to spermatozoa. In these cases the value of AIH becomes more debatable and results are sometimes disappointing. Where a husband is producing few spermatozoa, consecutive ejaculates may be freeze-stored in liquid nitrogen and then thawed and combined for use together at the appropriate time of the wife's menstrual cycle.

The important point about AIH is that by definition it involves a couple who are married. There are no obvious social issues to deal with. A husband and wife, we may say, are being helped to have a child and that is the end of it! But it is not as simple as that. At the biological level it could be argued that those who possess characteristics which prevent them reproducing in a 'natural' way are being allowed to do so by 'artificial' means. Whereas previously, the sub-fertile characteristics of such people were being naturally 'bred-out' of the population, they are now being retained and may be passed on to future generations, perhaps making sub-fertility and any other associated conditions more commonplace. If a greater need for sub-fertility treatment occurs as a result of the help being given to relatively few couples in this century, the social implications of present practice for the next century does have some meaning. There are echoes of the natural selection debates of the last century here and also reverberations of the more uncomfortable eugenic arguments that are still sometimes heard today. These arguments are relevant to all forms of artificial reproduction and they are difficult to deal with. Suffice it to say that the very word 'artificial' denotes an interference with the natural order; there is little doubt that eugenic implications are clearly present in some forms of artificial reproduction but its relevance to AIH is debatable. It is our view that the incidence of sub-fertility in couples leading to the need for AIH is sufficiently uncommon for it not to pose a serious threat to future generations, but the arguments are not without relevance.

Neither should it be assumed that there are no psychological issues surrounding the creation of a child by AIH. It is true the resulting child is that of the husband and wife and this removes many of the difficult social issues surrounding some other forms of artificial reproduction, but the relationship between a husband who has required help to produce a conception and a wife who may have no reproductive deficiency may still, on occasions, create tension. Much will depend on the reasons for the need to undergo AIH. The stress surrounding impotence is of a different order to that associated with the production of sub-fertile semen.

To our knowledge no study of the long-term effect of AIH on the marital relationship, or on the relationship of the parents to the resulting child, has ever been undertaken. The difficulties present in attempting to set up such a study are obvious but the information obtained, if sensitively collected, would be of considerable help in understanding some of the issues present in the more extreme forms of artificial reproduction.

AIH which is used to treat infertility has a comparatively low rate of success compared with AID where semen from a fertile donor is used rather than semen from a sub-fertile husband. The step from AIH to AID is an obvious and logical step in the treatment of some cases of infertility. From a purely medical point of view it may appear a small step, but the difference in the psychological and social implications of the two procedures is immense. Some practitioners have sought a compromise by offering a combination of AIH and AID. This compromise will be examined after a brief consideration of AID.

ARTIFICIAL INSEMINATION USING DONOR SEMEN (AID)

This form of artificial reproduction follows a similar procedure to AIH with the one difference that the semen placed within the wife is not that of her husband. No marital relationship exists between the donor and the woman receiving the sperm.

The first recorded instance of AID took place in America in 1884, but it was not until the 1930s, when advancing medical knowledge showed that in a considerable number of childless marriages it was the husband who was the infertile or sub-fertile partner, that the possibility of artificial human insemination began seriously to be discussed in Britain. AID could also provide a means of producing a healthy baby for couples where the husband suffered from an hereditary disease, or for couples who were unable to complete their families because of Rhesus factor incompatibility.

A small group of gynaecologists began to use the technique of AID during the Second World War, but their activities were virtually unknown to the general public until the first report of their work was published in the *British Medical Journal* in 1945. This report provoked a considerable amount of discussion, both in parliament and in the press, and the Archbishop of Canterbury instituted a commission of inquiry which recommended in 1948 that AID should be made a criminal offence. Following further debate the government appointed an interdepartmental committee, under the chairmanship of Lord Feversham, to make recommendations about the practice of AID (Feversham, 1960). They also concluded that AID was undesirable and not to be encouraged, but they feared that if it were made a criminal offence it would merely be driven underground and into the hands of unqualified practitioners. However, the committee felt that as the number of couples seeking AID was small, and the practice was being carried out discreetly by private medically-qualified practitioners, it was probably best left unregulated. Despite these discouraging noises AID continued to be practised and as infertile couples became more aware of this possible solution to their problems the demand for AID steadily increased.

11

Infertile couples had traditionally looked to adoption to relieve their childlessness but several developments during the 1960s meant that the number of babies available for adoption was drastically reduced; perhaps the most important of these was the Abortion Act of 1967. Attitudes to unmarried mothers also began to change and more single women decided to keep their babies rather than go through the trauma of relinquishing a baby for adoption. Increased job opportunities for women, and greater welfare provision also meant that single parents were financially somewhat better able to support their children. For infertile couples where the wife was fertile and able to have a baby of her own, AID now provided an attractive alternative to what often turned out to be frustrating attempts to adopt. By 1970 the British Medical Association was receiving sufficiently large numbers of inquiries about AID to warrant them appointing a panel of inquiry under the chairmanship of Sir John Peel. Peel reported more favourably (Peel, 1973) and recommended that AID should be made available on the NHS.

Since then a few NHS centres have been set up, but provision is still patchy with large areas of the country having no AID service, and in areas where clinics have been set up the waiting lists are often very long.

The Peel Report published in 1973 estimated that there would be a ceiling of approximately 1,400 married childless couples a year seeking AID. This natural ceiling ensured that numbers would always remain small, but again developments occurred which meant that this estimate was soon no longer realistic. Increasing numbers of couples who wanted no more children were turning to voluntary sterilisation as a means of more certain contraception. Increasing numbers of couples were also divorcing and remarrying, and so in the early 1970s a new indication for AID emerged. Men who had undergone vasectomy following the birth of the children of their first marriage were remarrying and wishing to father

12

children in their second marriage also. Reversal of vasectomy is rarely successful, adoption societies do not always look favourably on couples who have experienced a marriage break-up, and the best hope for many of these newly-married couples lay in AID.

A further development was that AID was now no longer restricted to provision within a stable marriage relationship. Within the social climate of greater freedom and self-determination for women, single women were also able to obtain AID from some practitioners in order to have a child without the necessity of a husband. Women living within lesbian relationships were also availing themselves of AID. Until the 1970s the number of AI practitioners was few and they were all, as far as is known, medically qualified. However, artificial insemination is a simple technique and some lesbian and single women began to realise that a medically qualified practitioner was unnecessary. A syringe bought from the chemist and the help of the (male) friend of a friend was often found to be sufficient.

The most important social implication of AID raised so far is the absence of a marital relationship between the man providing the sperm and the woman who provides the ovum. The product of their individual contributions is the child who, whether we find the thought distasteful or not, cannot ever be in exactly the same situation as most other children. The fact that the donor is not married to the mother throws up a whole series of social and personal issues which are described later in this book. Nevertheless, most AID is undertaken within marriage. But once a stable marriage relationship is no longer a necessary precondition for AID then the social and psychological implications of babies being born in households where no males are present, have seriously to be considered. This is not to say that single or cohabiting women or lesbian couples should necessarily be denied the right to have a baby by AID, but merely to point out the need for rules within which the service is to be provided if the practice is not to change what many see as the basis of our social organisation, the family.

The social and psychological implications of AID are important for yet another reason. Criticisms about the implications of newer artificial reproduction techniques such as ovum storage and donation, or surrogate motherhood, are often countered by reference to the practice of AID which, it is claimed, has already been taking place for decades without complaint or regulation. It is our belief that the key to the understanding of other forms of artificial reproduction and their regulation does lie within AID, but reference to the social acceptance of AID as support for other forms of artificial reproduction is absurd. The reason why the social implications of AID have not been discussed in previous years is not because of social acceptance of the procedure but because of the almost total secrecy which has surrounded the topic. This has hindered a social and psychological assessment of the procedure despite regular public calls for such an assessment since the technique was first reported in the *British Medical Journal* almost 40 years ago.

ARTIFICIAL INSEMINATION USING COMBINED HUSBAND AND DONOR SEMEN

Having looked briefly at the social implications of AID, we should now take a step back to AIH and consider those cases where mixing of donor's and husband's semen takes place. This is still practised and discussed but it is probably less common now than it once was, mainly because the addition of sub-fertile or infertile semen reduces the fertilising potential of donor semen. The donor semen is, as it were, diluted and the chances of conception occurring are reduced. Practitioners are therefore reluctant to mix semen in this way because the likelihood of achieving a pregnancy is reduced. But there are psychological and moral objections to this practice also. Presumably, the purpose in mixing semen in this way is to support the husband (and his wife) in the belief that the sperm which reached his wife's ovum to produce a child *could* have been his own. This may be seen as providing both a psychological and a legal support for the husband; but this support would be dishonest and could not be defended by any rational analysis of the situation. When a husband has tried without success, sometimes over a period of many years, to father a child and has been told that the chance of conception without recourse to other procedures is unlikely, to look then upon a pregnancy achieved following artificial insemination with combined donor's and husband's semen as though it were the husband's own spermatozoa which had fertilised the ovum, is neither rational nor honest. Blurring the issue, by combining semen, merely encourages a defensive denial of male infertility at a time when the husband is in greatest need of help in facing, and coming to terms with his condition. The experience of infertility will be explored at greater length later in this book, but for many men the knowledge of their infertility comes as a devastating blow which strikes at the heart of their feelings about their own worth as an individual. The psychological response of denial to such unpleasant news is a normal initial response, but surely the husband must be helped to work through his feelings of denial, anger and depression towards an eventual constructive acceptance of his infertility. Practices such as the mixing of semen, which purposely confuses the situation, merely encourage and prolong the initial response of denial and hinder the eventual constructive resolution of the conflict.

ARTIFICIAL REPRODUCTION TECHNIQUES FOR THE ALLEVIATION OF FEMALE INFERTILITY

Up until this point we have only considered techniques of

artificial reproduction which have been developed to overcome cases mainly of male infertility. In fact it would be more accurate to say that AID overcomes *childlessness*, as the condition of infertility is only circumvented and not cured. The causes of male infertility are poorly understood and treatment using drugs or other forms of therapy is rarely successful. Female infertility on the other hand is better understood and a wider range of treatment may successfully be used to bring about a pregnancy. Nevertheless not all cases are successfully resolved and recently techniques of artificial reproduction have been used to overcome female infertility also. Technologically the most simple procedure has been the use of artificial insemination combined with surrogate motherhood.

THE SURROGATE MOTHER

A surrogate mother comes to an agreement with another unrelated person or persons to conceive and carry a child which she promises to hand over to these persons once the child is born. Usually a couple, where the wife is infertile, contract with a woman who agrees to be impregnated using artificial insemination techniques with the sperm of the fertile husband. For a fee the woman is willing to donate an ovum and undertake the forty-week pregnancy. She agrees to hand over the child to the couple as soon as it is born. In some states in the USA legal contracts between the couple and the surrogate mother are drawn up and the appropriate fee agreed. The surrogate mother agrees to do all she can to remain fit during the pregnancy as part of this contract and guarantees to give up the baby once it is born. The only behaviour that has been avoided in this procedure is the act of sexual intercourse between the husband and the woman.

It is not difficult to see that at one level this is a form of AID working in the reverse direction. Instead of the sterile husband obtaining the help of a fertile sperm donor to assist his fertile wife to achieve a pregnancy, the fertile husband is using the assistance of a fertile woman outside his marriage to produce a baby his sterile wife cannot conceive. This form of surrogate motherhood involves many of the problems associated with AID. These relate mainly to the definition of parentage and the associated rights and duties of those concerned, but there is one significant feature which aggravates these problems. Whereas in AID the sperm is obtained from the donor and used independently of that donor, the surrogate mother not only produces the ovum which plays such an important part in the creation of the child, but she also carries and protects and nourishes the developing baby within her own body for a period of approximately forty weeks. The surrogate mother is genetically, physically, psychologically and socially involved in the creation and development of the growing child in a way that no male semen donor ever is. In fact it has been suggested that the motherhood of this woman is such a reality that the term 'surrogate' may be more appropriately applied to the infertile wife who hopes to take over the nurturing of the child once it is born. It is for this reason that where the surrogate mother has changed her mind and declared her intention to keep the child, even after receiving payment for the part she has played, social and legal support is usually given to her. Where an appeal to law has been made, in America and in England, to redress a breach of contract between a childless couple and a surrogate mother, the courts have found for the surrogate mother.

In both surrogate motherhood and AID there is a separation of genetic and social parenthood but the relative balance is different in the two situations. In AID social fatherhood begins immediately the wife's pregnancy is confirmed. One mother, pregnant following AID, wrote 'My husband has slept by the growing baby, felt it kick him, gone to classes in preparation for and longs to be at the birth of what is to all intents and purposes *our* child'. But this situation is reversed in surrogate motherhood; the childless couple merely produce some of the genetic material and they have no role to play during the important process of pregnancy. This is in contrast to the donor mother who spends nine long months physically and emotionally attached to the developing baby.

The question of sexual intercourse and its relationship with reproduction is again raised in such a case. Had the fertile husband sought a child outside his marriage through the means of sexual intercourse, few people would consider that the woman should be classed as a surrogate mother. Descriptions using words like 'mistress' or 'adultery' would be more likely and the integrity of the husband (and indirectly perhaps even his wife) would be called into question. By using means that avoid the need for sexual intercourse it seems that the opprobrium which would otherwise be attached to the whole procedure can be reduced. There have doubtless been cases where a childless couple have been helped by arranged sexual intercourse with a man outside the marriage or by the husband and another woman who gives up the child at its birth, but these are rarely reported. The very fact that when such a situation does take place it is undertaken with the utmost discretion and secrecy indicates an assumed social disapproval of the practice. It is the removal of sexual intercourse from reproduction that has made what was previously unacceptable somewhat more acceptable. We will return to this issue again.

This relatively straightforward type of surrogate motherhood still leaves unresolved the problem of who *is* the mother of the resulting child, but recent technological developments have produced additional and less straightforward possibilities. The arrival of external human fertilisation has led to other more complex forms of surrogate motherhood which have little in common with its male AID counterpart. We shall return to these following a discussion of external human fertilisation.

EXTERNAL HUMAN FERTILISATION

For internal fertilisation to take place the sperm and the ovum must be able to traverse, and meet, in the fallopian tube or oviduct. If the oviduct is malformed or malfunctioning internal fertilisation may be impossible, but in 1978 the first human baby to be conceived as a result of external fertilisation was born. Dr Robert Edwards and Mr Patrick Steptoe had given new hope to infertile couples who could produce normal ova and sperm but who, for a variety of reasons, could not provide the opportunity for fertilisation to take place.

Put simply, this procedure entails the collection of a mature ovum as it is released from the ovary. This ovum is placed in a special culture medium in a laboratory glass dish (hence the term 'in-vitro' fertilisation – literally 'in glass') and the husband's sperm is then added. If successful, fertilisation takes place and the early development of the embryo begins. The embryo is then returned to the mother's uterus ready for implantation in the usual way. It is clear that an immense amount of skill is required at all these stages – the collection of the mature egg, its fertilisation outside the human body and its return to the uterus where it develops as an embryo. As with AIH there is nothing in this procedure, involving a husband and wife, which directly raises any legal or social concern. By the use of advanced technology and medical expertise a childless couple has been assisted to have a child of their own and the scientific skills employed can only be applauded.

The possibility of producing a pregnancy without the need for sexual intercourse has been known and practised for decades, but until recently the ovum was always within the woman who was inseminated and was fertilised within her own fallopian tube. There was no doubt that the resulting child was hers even if there was uncertainty about whose sperm might have contributed to the other half of the germ cells which developed into the child. But with time the

19

ability to fertilise a human ovum outside the human body has become a reality. The assumption that the fertilised ovum which is replaced is indeed the woman's own ovum must now be taken on trust, and similarly that the sperm used to fertilise the ovum is that of her husband.

The assumption of marriage between the provider of the ovum and the provider of the semen is the important issue here, for whilst such a relationship exists the issue for society as a whole is a minor one. As with AIH, the genetic considerations of this procedure are largely academic. But sadly, it does not stop here, for whatever is possible for the married couple is also possible for anyone else. The combination of possible partners and their relationships is complex. We have already noted the separation of genetic and social parenthood implicit in the technique of artificial insemination by donor, but with the development of external human fertilisation the possibilities for the separation of genetic and social parenthood become far more complex. With artificial insemination the woman producing the ovum also sustains the pregnancy and cares for the child once it is born. Following external fertilisation the woman who sustains the pregnancy and gives birth may be a different woman from the one who provided the ovum, and different again from the woman who rears the child. We are back again in the realm of surrogate motherhood.

For couples who are childless because the wife, although able to conceive, habitually miscarries and is unable to sustain a pregnancy to term, the advantages of a womb-leasing type of surrogate motherhood are obvious. To be able to transfer the fertilised egg of the husband and wife to a secure womb before the danger of miscarriage occurs appears an attractive possibility. One could also imagine other situations where it would be medically hazardous for a woman to undergo pregnancy and advantageous for another woman to sustain the pregnancy. However, accompanying this apparently helpful procedure there are issues that are very difficult to resolve. Who is the mother of a child that is conceived by one woman but carried by another? We have already noted that there is an acceptance that the woman who both contributes her own ovum and also provides the early nurturance for the developing child possesses parental rights. But here we have a situation where the surrogate mother merely provides an incubation facility for a potential child that has been conceived by someone else.

Clearly, this type of surrogate mother is qualitatively very different from the mother who not only undergoes the pregnancy but also provides the necessary genetic material. Embryo transfer is not a matter of conjecture or science fiction; as this is being written a case of embryo transfer from one woman to another is being reported in the American press.

There are other, more futuristic, possibilities inherent in this development. With the help of external human fertilisation and surrogate motherhood it is now possible for women who are capable of conceiving and carrying a child to reproduce without the need to undergo a nine months period of pregnancy. Women in demanding occupations who would find a period of pregnancy a hindrance to their careers, or other women who do not otherwise wish to experience the disadvantages of the pregnant state, could employ another woman to fulfil this function for them. The developmental, social, psychological and legal implications of such an arrangement have no precedent. There is much more to the development of a child than just its genetic input. Some of the negative influences on the growing embryo are well known, for example rubella, alcohol consumption and smoking, but factors contributing a positive influence are more difficult to identify.

The desire to forego pregnancy in such cases must also be seen from the perspective of the woman who is to provide the surrogate facility. What factors would influence a woman to offer to carry another woman's baby and

21

relinquish it at birth? She may be motivated in an altruistic way, but it is likely that in many cases financial considerations would play a major part. Over many centuries economic necessity has caused some women to sell their bodies; this would bring a new meaning to the term.

When considering the motivation of ova or womb donors we should not forget the motivation of the semen donor who provides the sperm for AID. Why are men willing to provide semen, over the use of which they have no control, and can have no knowledge of the outcome of such use? It can hardly be for financial gain as monies paid are relatively small – though maybe £10 is not to be sneezed at if one is an impecunious medical student. Are these men motivated by sympathy for the plight of childless wives? Do they find the thought of their potential to beget children pleasurable? The men most commonly used as donors are medical students or the husbands of successfully treated infertile wives. Both these categories of men are to some extent beholden to those asking them to donate semen and may well feel that it is difficult to refuse.

The new possibilities which surrogate motherhood opens up also extend into the dimension of time. Theoretically one couple could conceive on a number of occasions in the same year and have each embryo grown by different surrogates! The question of time is important in another respect which is of far more direct relevance, for it has been practised for many years and could be described as a growth industry in the 1980s. This relates to the ability to freeze-store semen and embryos.

CRYOPRESERVATION OF SEMEN AND EMBRYOS

The freeze-storage of sperm is not new but only recently has the freezing and storing of embryos taken place. This means that a child conceived today outside the human body using external fertilisation techniques could be stored in an 'embryo bank' until it is required for development. The *timing* of conception and birth will therefore not necessarily be associated in the usual way. The duration of the pregnancy will, in a sense, be prolonged by the inactive state induced by the freezing of the early embryo. Birth need not inexorably occur forty weeks or so after fertilisation but may take place at the convenience of human decision, perhaps several years later.

In itself, the period of approximately forty weeks between the act of conception and birth of the child may not, at first, appear to be important – at least in terms of its social significance. But closer scrutiny raises issues which reveal the unquestioning assumptions which underlie much of our social life. When a child is born about forty weeks after an act of sexual intercourse, it is assumed that those responsible for the creation of the child are also those who engaged in the earlier sexual activity. Normally, the claim made that a child born many weeks after the forty-week interval has been fathered by a person who admits to an act of sexual inter-course but denies paternity, is usually discounted. There comes a point at which the time interval is so great that it is logical to assume that a particular act of sexual intercourse is not connected to the conception of a particular child. With the arrival of artificial reproduction techniques and the ability to store sperm and embryos, such certainty is removed. Here is an example of the separation of repro-duction from sexual intercourse having an indirect affect on the relationship between the act of conception and the birth of the resulting child.

By removing sexual intercourse and internal fertilisation from the equation, the certainty that a child is 'his' or 'hers' by calculation of retrospective dates and behaviour is no longer present. The introduction of cryopreservation as a technique associated with the practice of artificial repro-duction has created a level of complexity into the practice

23

which is, at first, difficult to realise. Cryopreservation can affect every form of artificial reproduction and is not confined to the most advanced or recent varieties associated with external fertilisation. We have to go back to the beginning, for the effect of this new technique even on AIH has social and psychological implications not previously discerned.

The technique of cryopreservation has now made it possible for a husband to deposit semen for long-term storage in a sperm bank. This is most commonly done before a man undergoes a vasectomy operation but the facility is available for anyone to use for a fee. It is therefore now possible for a wife to be impregnated with her husband's semen after his death and for his child to be conceived posthumously. This, as some might think, ghoulish procedure has already been undertaken. Presumably the husband and wife had previously discussed the possibility and come to an agreement about it. However, the resulting child, arguably the person most affected by the decision, was in no position to make his or her point of view known. If we are to consider the rights and needs of the child, the wisdom of such a procedure requires careful consideration. What is the status of a child created from the sperm of the now-dead husband of the child's mother? Marriage ceases at the death of one of the partners and the question of legitimacy therefore becomes an important one.

In the case of AID, selected semen can be stored for lengths of time previously not possible. Furthermore such semen can be transported for storage and used anywhere where cryopreservation facilities are available. Not only have the time boundaries been removed in relation to reproduction but the geographical boundaries which previously placed considerable limitation upon who reproduces with whom have also been removed. The social implications of the changes introduced by the possibility of freeze-storage have hardly been considered.

External fertilisation raises yet more profound questions. The technique of stimulating the ovary of an infertile woman to develop more than one ovum at a time is now relatively commonplace. Where childlessness is being treated by external fertilisation techniques there may be a higher rate of success if several ova are collected on one occasion and attempts made to fertilise each of them. Where fertilisation is successful, one or two of the fertilised ova are replaced in the uterus for implantation and development in the usual way. To replace three may be counterproductive as there is a greater likelihood of complications arising in a triplet pregnancy. How then are any spare embryos to be dealt with? There are three possibilities: the first is simply to destroy those that are not replaced in the uterus. The second is to use this material in order to learn more about the beginnings of human life through a process of laboratory experimentation. The third is to freeze-store the 'spare' embryos for subsequent use either by the woman from whom the ova were collected or some other woman. All these options raise serious and far-reaching questions. The social implications of artificial reproduction at this point must include an assessment of the moral and ethical issues surrounding the sanctity of human life and the time at which such life begins. It would be salutary to remember here that all the techniques discussed in this chapter, AIH, AID, external fertilisation and embryo transfer have as their desired outcome the creation of a child. In making decisions and undertaking procedures to fulfil the wishes of would-be parents it would be improper to forget that some child must live with the implications of those decisions and procedures.

2
A Necessary Vocabulary

The discerning reader will have noticed already in the previous chapter that the use of some words and terms was inappropriate because uncertainty surrounded their precise meaning. The most obvious example is that of the term 'surrogate motherhood' which is used to describe many different situations. The difficulty stems largely from the novel development whereby the use of the new techniques that have been described in Chapter 1 permits the process of reproduction to be broken down into separate phases or processes. The fusion of germ-cells in fertilisation, the implantation and subsequent development of the embryo during pregnancy, and eventual child-rearing, can now be seen as discrete events or stages in the reproductive process, each of which may be undertaken by a different person. The vocabulary of the English language does not provide words to describe accurately the roles played by these different persons who carry out a particular phase of the reproductive process. The relationship in which these people stand in respect of each other also adds to the confusion. It is no longer always the case of a father and a mother, who are husband and wife, being parents.

It follows that it is not sufficient to discuss the techniques of artificial reproduction in isolation; these techniques must be understood in terms of the social situation in which they are used. External fertilisation of the ovum of a wife by the sperm of a husband, both of whom are to undertake the bearing and rearing of the subsequent child, may be using exactly the same laboratory techniques as the external ferti- lisation of the ovum and sperm of individuals who may not know each other, and who will know nothing of the subse- quent pregnancy or child-rearing, but the social and psycho- logical implications of those same laboratory techniques are very different.

If we are to explore and define the social implications of the new technology we need an adequate vocabulary. Words are necessary for understanding and communication; they clothe and encapsulate concepts and make possible shared meanings. If the words we use describe more precisely the social situation in which the newly available techniques are used, then it becomes possible to determine the potential areas of confusion or difficulty more easily. The need for careful planning and institution of controls can then be more sympathetically demonstrated.

Thus at this point we should pause in order to equip ourselves with a vocabulary or terminology which describes more accurately the *techniques* entailed in artificial repro- duction, the *roles* played by the people involved in the processes of reproduction, and the *relationships* between these people. Each of these three types of definition (tech- nique, role and relationship) depends upon some under- standing of the other two and cannot be considered totally in isolation from them. For example, in most cases of AID described in Chapter 1, the technique of obtaining semen from a person performing the role of semen donor is pre- sumed to be undertaken to assist a woman who is related by marriage to a man who cannot produce fertile semen himself. To appreciate the social implications of AID we need to know something about each of these three constituent elements in the situation, for there are occasions when an alternative scenario to this example may be followed but which is still called AID.

TECHNIQUES OF ARTIFICIAL REPRODUCTION

ARTIFICIAL REPRODUCTION

This term describes all techniques whereby a conception is effected by means other than by sexual intercourse. Notice that this definition says nothing about the relationship between the woman producing the ovum and the man producing the sperm. Nor does it indicate when and where fertilisation takes place. In artificial reproduction fertilisation can occur at any time, even years after the production of the sperm; it can also take place inside or independently of a woman's body. Artificial reproduction techniques may be simple to use, perhaps requiring no more than a syringe used by the woman desiring a pregnancy or they may be highly complex requiring the assistance of technical and gynaecological expertise. The important issue is that reproduction is effected by means other than sexual intercourse. There are two forms of artificial reproduction depending on where fertilisation takes place – internal or external to the woman's body. We call these 'artificial insemination' and 'external human fertilisation' respectively.

Artificial insemination
This form of artificial reproduction occurs when the woman providing the ovum has that ovum fertilised whilst it is still within her own body. The woman who produces the ovum also produces the environment in which it is fertilised naturally, using semen which has been artificially introduced. AIH describes the situation where the semen of the woman's husband is artificially introduced and AID refers to the same procedure but this time using the semen of a man who is not

28

related by marriage to the woman concerned. These procedures were described in Chapter 1.

External human fertilisation

Most people would recognise this form of artificial reproduction as '*in-vitro* fertilisation', or more popularly as the creation of a 'test-tube' baby, but the choice of an alternative term to describe this procedure is deliberate. The term '*in-vitro* fertilisation' describes a procedure in technical language which avoids reference to the human beings concerned. The use of such a term tends to focus our attention on the procedure that is taking place and so reduces the stimulus to study the people involved, including the child yet to be born. The term 'external human fertilisation' places emphasis on people.

Embryo replacement and embryo transfer   Once the ovum has been successfully externally fertilised, the early embryo must be introduced into the uterus so that it can implant on to the uterine wall and continue its pre-natal development. Previously this procedure was called 'embryo transfer'; the early embryo was transferred from its culture medium to the uterus. More recently this term has been re-defined. If the embryo is returned to the uterus of the same woman from whom the ovum was obtained, the procedure is now known as *embryo replacement*. Only if the embryo is introduced into the uterus of a different woman, a woman who did not produce the original ovum, is the procedure now (more accurately) defined as *embryo transfer*.

An alternative form of embryo transfer has recently been reported. In this procedure a female donor is artificially inseminated with semen from the fertile husband of an infertile wife. The resulting embryo is then flushed from the uterus of the donor and transferred to the uterus of the wife who then carries and gives birth to the child.

ROLES IN ARTIFICIAL REPRODUCTION

The distinction between the procedures of artificial insemination and external human fertilisation is important for it provides the first step in identifying the different phases in the reproductive process, phases which may now be undertaken by different people. It is now possible to separate out the reproductive processes which traditionally have been considered as a single process called 'motherhood'. The processes involved in the production of an ovum and the development of the embryo after the ovum has been fertilised have been inexorably linked to the same woman since human reproduction began. Through the technique of external human fertilisation the separation of these reproductive processes is now possible. The woman who produces the ovum need no longer be the woman who grows the resulting embryo. It is this separation of reproductive processes that creates so much difficulty when it comes to an examination of the social implications of artificial reproduction. The separation of reproduction from sexual intercourse has now been joined by the possible separation of ovum production and pregnancy.

So far we have discussed the reproduction in terms of identifiable biological processes, two of which have been linked since human reproduction began but which can now be separated. To permit a social perspective, as distinct from a biological one, these processes must be translated into behaviour of some sort. The social implications of reproduction depend upon the identification of appropriate (and inappropriate) behaviour in a given social situation. For convenience, as a means of discussing something we all understand but find difficult to describe, social scientists have introduced the concept of 'roles'. The set of behaviours which describe being a son, daughter, wife, husband, mother or father are each called a role and most of us would find little difficulty in describing in broad terms the kinds of behaviour appropriate to each of these roles. For a woman to be called a 'mother', for example, we would usually expect her to be responsible for the production of the ovum which provides half the germ cells of the child, the pregnancy and birth of the child and the care of the child after its birth. In the past the mother role encompassed all these things and where departures from it occurred, as in the case of the fostering or adoption of a child, special arrangements were made using procedures developed and openly agreed upon within our society as a whole. Even the identification of the mother role as we have described it does little justice to the complexity surrounding what we really mean when we attempt to describe what being a 'mother' entails, but the complexities introduced into this role by the use of some artificial reproductive techniques cannot be exaggerated.

An understanding of the broadly defined roles involved in human reproduction brings into sharper focus the roles individuals play in the procedures of artificial reproduction. Once a role is discerned, the justification for examining its effect upon society becomes more apparent. In the usual situation the male who produces and delivers the sperm and who cares for the resulting child after its birth is the same person, and fulfils the role of 'father'. Similarly the three processes requiring a female, that is the production and delivery of the ovum, the protection and development of the embryo and foetus once fertilisation has occurred, and the nurturing of the child once born are also usually performed by the same person fulfilling the role of 'mother'. But with the introduction of artificial reproduction it is possible for each of these processes, both on the male and the female side, to be performed by different people; that is, up to a total of five people in different combinations instead of the usual two. Each combination introduces differing social issues which may be important for family life and thereby for society at large. The use of descriptions such as the role of 'mother' or 'father' in such circumstances gives a false

31

Artificial Reproduction

impression of the role actually played by each person. Some redefinition of these roles is necessary but there are currently no words to describe the alternative functions of parenthood associated with these separated roles. The need for a nomenclature is clear, for depending on the definitions and descriptions used it will only then be possible to define more closely the social duties and obligations of those directly concerned.

A SUGGESTED NOMENCLATURE

The three processes which contribute to reproduction in the female are described in terms of the woman's role in the process.

a) *The genetic mother*. This describes the role of the woman who produces and matures the ovum or egg. The term genetic mother is preferred to the alternative 'biological mother' when describing this role because it is more precise; to label this role 'biological' would not differentiate it from the role played during pregnancy.

b) *The carrying mother*. This refers to the role of the woman in whose uterus the embryo implants and develops into the growing foetus. Again, the alternative of 'pregnant mother' has been discarded in favour of this more precise term. The term 'pregnant' describes a current condition and to call a carrying mother a 'pregnant mother' when the carrying took place some time previously creates semantic difficulties.

c) *The nurturing mother*. This term refers to the role of the woman who will care for the baby once it is born, and is preferred to the latin term *mater* which could be confused with the genetic and carrying mother.

d) *The complete mother*. So far we have three types of mother: the genetic mother, the carrying mother and the nurturing mother, each fulfilling a particular role in reproduction. Where the same woman undertakes all

32

three roles, the term 'complete mother' would apply. This is not meant to indicate the quality of the mothering but simply the fulfilment of all three roles normally associated with motherhood; provision of the ovum, pre-natal development and the nurture of the child once born.

e) *The genetic-carrying mother.* This mother supplies the ovum and protects the child during its pre-natal development but does not keep the child once it is born. This term describes the usual case of 'surrogate-mothering' as presented in the media but it also covers those situations where an unwanted conception occurs and the baby is given up for adoption soon after its birth.

f) *The genetic-nurturing mother.* This term describes the role of a woman who has produced the ovum and who wishes to nurture the child but for some reason cannot, or does not wish to go through the forty-week pregnancy. There may be medical reasons which indicate a need for a carrying mother but there may also be economic, professional or psychological reasons which may have nothing to do with a woman's physical inability to experience the pregnancy.

g) *The carrying-nurturing mother.* This term describes the role of the woman who carries the child during pregnancy and cares for it after birth but who did not provide the ovum. The situation is the female version of AID and would presumably be indicated in similar circumstances, that is, when ovum production is absent or unsatisfactory, and in cases of genetic abnormality. This situation would also occur if successfully externally-fertilised 'spare' ova from one woman were implanted in the uterus of another woman for whom external fertilisation of her own ova had failed.

The most obvious question that arises following the description of these seven possible situations is, 'Who is the

33

Table 2.1 *Possible Categories of Motherhood and Fatherhood as a Result of Artificial Reproduction*

Genetic mother
Carrying mother
Nurturing mother
Complete mother (combining the genetic, carrying and
    nurturing roles)
Genetic/carrying mother
Genetic/nurturing mother
Carrying/nurturing mother
Genetic father
Nurturing father
Complete father (combining genetic and nurturing roles)

---

mother?' This seems a natural question to ask but it is one for which there is no answer at the present time. New rules are required where none have existed previously and where no precedents are available. The adoption experience is of little help because the separation of genetic and carrying motherhood is not present. The belief that during early life the relationship between the carrying mother and the developing foetus is of a very special kind serves only to make the situation more complicated. What rights have the genetic, carrying and nurturing mothers in relation to the child and, perhaps more importantly, what rights and what needs does the resulting child have in relation to each? Can a signature on a form disclaiming rights by the genetic mother also remove the possible rights and needs of the child? Once we begin to ponder the issues surrounding motherhood the realisation of what we have taken for granted for so many generations raises questions which our present legal and social system is ill prepared to answer. If artificial reproduction is to continue, and there is little doubt that it will, then some acknowledgement of the rights and responsibilities accruing to each of the different categories of 'mother' have to be made explicit.

Yet this is not the end of the matter. Up until this point consideration has been confined to the female; the male also has a part to play. He is involved in just two processes, that of the production and delivery of the sperm and the nurture of the child after its birth. This restriction to two roles reduces the complexity in terms of the number of role combinations when compared to the female, but many of the qualitative differences remain.

h) *The genetic father.* This describes the role of the person who provides and delivers the sperm for either internal or external fertilisation.

i) *The nurturing father.* This description defines the role of the man who cares for the child after its birth but who was not directly responsible for its creation. The nurturing father is assumed to be the husband of the woman who experienced AID, or who has adopted a child or who for any other reason has taken on the responsibility of nurturing a child which is not genetically descended from himself.

j) *The complete father.* Where the roles of genetic and nurturing father are performed by the same man, then the description of complete father is appropriate. Again, this term is not meant to indicate the quality of the fathering.

Having differentiated three types of fatherhood – the genetic, nurturing and complete father – the same question arises as for the seven classes of motherhood, who is the 'father' in terms of the responsibilities we normally associate with fatherhood? The complete father like the complete mother provides few problems, but what of the genetic father and the nurturing father? In cases of AID the genetic father would be the donor and the nurturing father is assumed to be the husband of the woman receiving the

35

donor's sperm. But what rights and responsibilities does each 'father' possess in relation to the other, and in relation to the child? Once again the larger question of the rights and needs of the child in relation to its parental background require clarification. We have now reached what was previously a science fiction situation in which it is possible for a child to have up to a total of five parents – that is, three 'mothers' and two 'fathers'.

RELATIONSHIPS IN ARTIFICIAL REPRODUCTION

Yet another dimension to be considered when seeking clearer social definitions of artificial reproduction is the relationship between the man and woman taking part in the procedure. There are two sorts of relationship to be considered; those linking the reproductive roles the man and the woman are fulfilling, and those providing a social association of a definable kind. The linking of the roles of the nurturing-father and genetic-father is an example of the first type and marriage is an example of the second. There is an assumption that the nurturing mother and nurturing father who share the responsibility for the care of the growing child will be related by marriage, but this remains only an assumption.

An attempt has been made to demonstrate the different roles and relationships in some of the various forms of artificial reproduction in Table 2.2. Some of these relationships are comparatively easy to identify and create little or no confusion. For example, where a complete mother and complete father combine as in AIH, they provide all the usual roles of child creation, gestation and nurturing between them. In embryo replacement information describing the mother's and father's marital status is missing; there is an assumption they are married to each other but this need not necessarily be so. As far as the techniques themselves are concerned, the need to discuss relationships may appear

irrelevant but nevertheless when most people describe artificial reproduction they assume relationships exist which
may, in reality, be absent. If marriage is considered to be
important to the relationship between mother and father
when describing artificial reproduction techniques, then yet
another set of terms, or at least a suitable prefix, would be
required to describe those situations where marriage is or is
not present.

The practice of surrogate motherhood following artificial
insemination illustrates yet another complexity; when
reproductive roles are separated it is necessary to consider
who is filling the unstated role. The use of the description
'surrogate mother' assumes that the woman to whom the
genetic-carrying mother gives up the child, the woman who
will fulfil the role of nurturing mother, is the wife of the
complete father. It is also worth noting that the decision to
describe the genetic-carrying mother as the surrogate or
substitute is an entirely arbitrary one; the term surrogate

Table 2.2   *Roles in Various Forms of Artificial Reproduction*

Artificial insemination	External human fertilisation	Roles
AIH	Embryo replacement	Complete mother and complete father
AID	—	Complete mother and nurturing father
—	Embryo transfer	Carrying/nurturing mother and complete or nurturing father
Surrogate motherhood	—	Complete father and nurturing mother (using the services of an unrelated genetic/carrying mother who is artificially inseminated by the father's sperm)

37

mother could equally well be used to describe the role of the woman who makes herself responsible for the nurture of the child. The assumption in describing the genetic-carrying mother as the surrogate seems to be that this is the lesser role compared with that of nurturing mother. But this is a large assumption; the bond between a mother and her unborn child is as yet not adequately understood, but it is quite evidently a bond of considerable strength. We will return to a discussion of the relative importance of the separated reproductive roles later.

We have seen how the complexity of artificial reproduction increases when social factors are considered. To describe external fertilisation merely in terms of technique is so simplistic that it has little relevance to the social world in which all the participants in artificial reproduction reside and into which they are hoping to introduce a new human being.

It seems that provided the nurturing parents are married and provided sexual intercourse is avoided between genetic parents who are not married, then artificial reproduction can be contemplated! Of course, this is an oversimplification but the disentangling of roles played by all those involved is dependent upon determining who is married to whom and who avoids having sexual intercourse with whom. This is not a facetious comment for the central positions of marriage and sexual intercourse will become more pronounced the more closely the social implications of artificial reproduction are examined.

A start has been made in providing a vocabulary which it is hoped will help to illustrate the social implications of artificial reproduction. In this chapter we have tried to create the means whereby these implications can best be understood. Before attempting to define social parameters for artificial reproduction, it is necessary first to examine how 'natural' reproduction has been controlled through the accepted standards and norms which govern reproductive behaviour in our society.

3

The Family

Most people living in our society would accept the proposition that sexual relationships, the birth of children and the nurture of infants and young children are normally contained within the set of relationships we call the family. The family is a social organisation and very closely related to the social institution we call marriage. This is not to deny that other arrangements are sometimes practised or that some people deliberately choose to ignore the institution of marriage, but in the normal situation marriage and the family are contiguous. To understand the social implications of artificial reproduction it is necessary to determine the extent to which these practices affect the social organisation of the family and the institution of marriage and in order to do this we have to ask what is meant by family and marriage.

We all have knowledge of families, indeed most of us were born into one and most of us spend a significant part of our lives interacting with members of our own families. Because of this close personal experience of family life we are apt to assume we know what we mean when we use the word 'family'. Moreover, we also assume that the word has the same meaning for us that it has for others, that our experience is similar to that of other people who are not members of our particular family, but members of other families. Sometimes we use expressions indicative of our subjective feelings about family relationships which, if we consider them dispassionately and with objectivity, make little sense. Thus we may think we know what is meant when someone uses the phrase 'blood is thicker than water', but on closer examination there is considerable uncertainty as to what is really meant. Does it mean strictly 'blood-relationship' or does it include kin related by marriage as well as by descent? In our own family, who would we include in this category and how far would it extend? Herein lies a difficulty facing those who want to examine the family objectively, for our own experience, together with a confusion of subjective feelings about our family relationships, has led to the making of many assumptions about family life. The consequence of this is the difficulty of discerning the principles on which family relationships depend. Thus it comes about that the word 'family' is one of which most people think they know the meaning but which tends to mean different things to different people. The word is even used in other contexts as, for example, when we say we are 'familiar' with a particular area of knowledge, meaning we are 'at home' with it, or we fully understand it. But the fact is we often don't fully understand the complexities of family organisation, we either merely believe we do or we are basing our claim on a part of what it is possible to understand.

TYPES OF FAMILY

Let us ask again: what do we usually mean when we refer to 'the family'? Is it a married couple, a couple with children, or three generations of related people? Does it include in-laws or not? Moreover, even were we to choose one of these as our definition, the passage of time brings alterations so that we are bound to ask if what we defined as the family at one time is still relevant some time later when births and deaths and marriages have taken place. Even if we use a very precise definition the fact remains that the content of the description changes over time. Furthermore, how does the family relate to the household? Usually we think of the family domiciled in one place within one dwelling, but clearly this is not always so, for children grow up and leave the home, grandparents may, or may not, live with or near their grown-up children, and so forth. The family is not a static, unchanging object that can be described once and for all but a set of rather special human relationships which are ever changing and come in various forms.

What would be helpful is a means of identifying different kinds of families rather than precisely defining the family itself. Thus we may distinguish between the family of origin or orientation, and the family of reproduction. The *family of orientation* denotes the family which brought up or raised a child to adulthood, whereas the *family of reproduction* describes that adult's own family resulting from marriage and the birth of his or her own children. The family of orientation will include a person's parents, his brothers and sisters; but it may also include other kinsfolk related to each parent. The family of reproduction is more narrowly defined to include only one's spouse and the children of the marriage.

It is possible to make further distinctions. Thus, if we accept the requirement that to be a family the group must include children, we can distinguish childless couples from those with children. Is the married couple without children to be denied the title of 'family', whereas the unmarried couple, single woman or man, or even a couple of the same sex, who have at least one child of their own living with them to be given this description? And what of the adopted or fostered child who is not genetically linked with his or her nurturing parents but who is living in the same household and for whom the adults admit to some responsibility? Thus, the description of what constitutes a family is not as simple and straightforward as is often believed. What we can say is that an important element of family life is the presence of one or more adults who have assumed responsibility for caring for the children who live with them. Usually there are two adults, one man and one woman, but occasionally only one adult is present. We call such adults 'parents' and usually they have a formally recognised relationship to each other provided by the institution of marriage. It is frequently asserted that the institution of marriage is becoming weaker, and is an outmoded concept, but there is little evidence to support this point of view. To be sure the incidence of cohabitation before marriage is increasing but when cohabiting couples decide to have children they almost invariably marry. Nor can the high incidence of divorce be used as evidence for a decline in the influence of the institution of marriage, for the incidence of remarriage after divorce is also very high. There may well be dissatisfaction with specific marriage partners but the high rate of remarriage shows that this dissatisfaction does not extend to the concept of marriage itself. A considerable proportion of children do live in one-parent families but it should be understood that this is not a static state of affairs. Children born to a mother who is unmarried tend to move into a two-parent family as the majority of single mothers eventually marry. Similarly a child living with a single divorced parent will have commenced life in a two-parent family and is likely to return to that situation as his or her parent remarries. The common incidence of one-parent families is sometimes used to justify the provision of artificial reproduction techniques to single women who wish to have a child but do not wish to marry. But the two circumstances are not analogous; the majority of children who live in a one-parent family have lived, or will in future live, within a two-parent family for some period of their development. Clearly, marriage has something to do with the family, but equally clearly something we may call a family may exist outside the bonds of matrimony. Suffice it to say at this point that marriage is a public declaration by two people that the offspring of the marriage are legally theirs and therefore part of 'their' family.

At least we may say that the term 'family' refers to a group of two people with their offspring and this is normally how we view it, but marriage is not essential except in the light of certain legal requirements. What then is essential? Is it a certain kind of relationship? In most cases there are objective relationships acknowledging paternity, motherhood, and so forth, and with these go subjective factors, namely the feelings and emotions which those who are related to each other have for each other. Such feelings prompt an attachment of the parents to each other which is exclusive and the most obvious manifestation of this exclusiveness relates to their sexual and reproductive relationship.

THE FAMILY AND REPRODUCTION

In the normal instance a man and a woman living together may expect to have a sexual relationship. We need not speak of 'rights' necessarily but certainly of expectations about each other which are mutually recognised. If the two are married then a certain formality is present which supports in law these expectations. Such expectations are not just in relation to the sexual gratification of the couple alone, they relate negatively to other relationships. In plain words most married couples require their sexual relationships to be exclusive. During the marriage, neither the man nor the woman is expected to have a sexual relationship with any other person. Indeed, adultery is strongly condemned and may well be grounds for a divorce of a married couple, or a breakdown in the relationship between the two. Indeed, one of the most common reasons for a disruption of a marriage and the disorganisation of a family is adultery; i.e., the breaking of an exclusive relationship. The very fact that childless couples where the husband is infertile choose to undergo the inconvenience, stress, embarrassment and financial strain of an impersonal and clinical procedure such as AID, rather than achieve a pregnancy by personal sexual intercourse with another man, serves to illustrate in quite a striking way the strength of the belief that sexual relationships within marriage must be exclusive. In addition to this exclusive sexual relationship there is the expectation that both parents will nurture any children of the marriage. This is a joint matter in which mutual comfort and support is provided. So we have two essential ingredients when describing family life: first, an exclusive sexual relationship, and secondly the birth, nurturance and upbringing of children. Family and marriage are thus concerned primarily with issues surrounding reproduction.

In the most common forms of family life the process of reproduction directly concerns all members of the family. The behaviour of each family member is identified by reference to the role each is expected to play. Sometimes the same person plays more than one role but each is sufficiently defined to allow family members and even those outside the family to identify the particular role being played. Thus we can speak of the roles of husband, wife, father, mother, daughter and son in terms that most people would recognise. But what does it mean to be a father, or a mother, or a son or daughter? We have indicated what it is to be a husband and a wife in terms of a sexual relationship but when we speak of father and mother attention shifts to the reproductive and nurturing functions of family life. Usually there is a division of labour based on sex differences and masculinity indicates some roles just as femininity indicates others. These roles are transmitted to the children from an early age. In the normal instance the children learn that sexual relationships between their parents are exclusive and hence they are children of the marriage. They are the genetic offspring of that exclusive relationship. Not only this, but the children of the marriage see each other as also resulting from the same exclusive relationship and hence as brothers or sisters to each other. The assumption of a genetic link is therefore both lineal in relation to one's parents and collateral in relation to one's siblings. This aspect of family life and upbringing rests on an unspoken assumption that a genetic link exists between brothers and sisters and their parents. There is a special case, namely adoption, where there is no genetic link, but then adoptive parents normally have made a public declaration of their status; it is made explicit. But let us note that the adopted children like the natural children of a marriage will also assume the exclusive nature of their adoptive parents' sexual relationship.

RELATIONSHIPS WITHIN THE FAMILY

Just as we have earlier indicated the peculiar use of the expression 'to be familiar with', so we point to the expression 'have relations with'. This latter expression of general utility is, of course, rooted in family life, for we speak of our 'relations', and by this we mean our kinsfolk. The set of kin relationships we find in family life is very important; it is unique and special; it is encrusted with privilege. We may be familiar in our dealings with parents and siblings and children, when we cannot be so easily familiar with others outside the family circle. It is not the case that there are necessarily very loving relationships within the family, although often there are, but rather that the expectations the members have of one another allow for a certain informality and ease of interaction. Moreover, the members of the family usually trust one another. Families have their secrets and keep them more often than not, but such secrets are shared by the family group. The family group is not confined to those living together in one household but extends to wider kin, although the quality of such relationships may vary considerably from one instance to another. Yet whilst this variation may exist the special kind of relationship persists – it is a family relationship. Thus kinsfolk may be called on for help, for support on the occasion of marriages and funerals, they may be relied on for comfort and encouragement, for overcoming family quarrels and for moral exhortations or discussion. What we are saying is that however we may define the family these special relationships are part of it. We can speculate on the reasons why these special relationships are so important. It may be that they contribute to family life by reducing tensions and providing the means for relaxing and 'being oneself'. On the other hand, the family may create tensions, albeit tensions which need to be generated in order to enable the members to achieve goals, to grow up and to achieve maturity. Certainly, we can see how the two-generation family provides protection for infant children; indeed the family is child-oriented. It appears to exist for the sake of the child although it persists after children have left home. But we may note how it is revivified when grandchildren appear and provide new roles for the older members of the family of orientation.

Whilst it is difficult to define the family, because it can be narrowly or extensively perceived and because it changes over time, we have been able to identify features of family life in terms of kinship relationships. We must now ask: on what basis do these relationships rest? Usually people use words like 'blood' and speak of 'blood relations', less frequently the word 'stock' is used and so people of the same stock are held to be related. What is meant here is that there is a genetic connection. Indeed, the small child learns at a very early stage that he or she *has* a mother, a father, brothers, sisters, grandfathers and grandmothers, uncles and aunts. The expression 'blood relations' sometimes almost smacks of a property relationship, as if people belong to one another. One is in a given situation with people to whom one is related; neighbours and friends may change but relatives never. Herein lies an important factor, for the familial relationships normally imply a shared genetic background. The genetic link may be direct as in the case of children and their parents or indirect where the birth of a child has a retrospective effect. For example, a married couple who are childless are not genetically linked, but when they have a child they each possess a genetic link with the child they jointly produce. Similarly, the relationships between the kin of the husband and the kin of the wife remain genetically separate until an indirect link is created by the birth of a child. Viewed in this way, reproduction becomes a central feature in the uniting of previously un-related sets of kin. Occasionally when a couple remain childless both the direct and indirect genetic links are absent but a close 'family' relationship is nevertheless maintained. Such relationships exist because a public declaration in the form of a recognised marriage ceremony has taken place which allows for the bringing together of two sets of kin who become related 'in-law' during the period of the marriage. Where children are born in the absence of marriage, relatives may or may not approve but most would be able to define their familial relationship with the child. The difficulty is found in defining the familial relationship with the parent of the child. If a sister has a child by a man with whom she is cohabiting but to whom she is not married, the relationship of aunt or uncle to the child is clear, but the family relationship with the sister's partner causes confusion. He is certainly not a brother-in-law and as a result the relationship of all his kin to one's own kin or oneself is confused. The childless marriage may not allow the closeness of a shared indirect link, by providing equal status to two sets of grandparents for example, but the public declaration of marriage removes any confusion about who is related to whom. It is worth reiterating at this point, that the central feature of marriage is the implicit assumption that the children of that marriage are those of the couple concerned who commit themselves to the care of these children. The importance of the related issues of reproduction and marriage are again emphasised.

It is the special relationship that the direct or indirect genetic link creates that gives a child and other relatives a right to have access to other close family members and to have knowledge about them. Hence a child has a 'right' to know who his genetic parents are if this is known. In the case of adoption this right is invoked in the advice to tell the child he or she is adopted, and legal provision has been made for a child, at a certain age, to have access to the original birth certificate identifying the genetic mother, and in some cases the genetic father, even though these parents abandoned their nurturing role in favour of adoptive parents. Of course, in the 'normal' family the child does not have to be told who his immediate relatives are; formal instruction is usually confined to identifying the more distant relatives who are met infrequently. That explicitness is *not* required does not detract from the importance of his assuming his genetic origin in personal terms. Whilst recognising the importance of genetic factors in the social interpretation of family life, it would be a mistake to assume that this was the sole feature. One's genetic background or genealogy may be objectively important in a social sense but we have already seen that the most subjective and personal issues associated with family life are of equal importance. Our perceived genetic links may help us to identify our kinsfolk but our feelings and sense of obligation and indeed the recognition of our own 'rights' in relation to these kinsfolk determine how we interact with them. These subjective interpretations of a special relationship with others are based on a number of assumptions and expectations which are seldom made explicit. The most obvious unspoken assumptions concern the relationship we hold with our parents.

In Chapter 2, special attention was paid to the three functions or phases of motherhood which, in the usual situation, are undertaken by the same woman. In defining the role of mother, most people would expect that the genesis of the ovum, its development once fertilised and the nurturing of the child are all undertaken by the same person. The father of the child is also expected to have been responsible for the genesis of the sperm which fertilised the ovum and to be willing to undertake joint responsibility for nurturing the child after its birth. Furthermore, most couples responsible for the creation of a child are expected to be contracted to each other by marriage. We have already noted that marriage implies exclusiveness in relation to sexual intercourse and this assumption is usually held by both married partners, the children of 'their' marriage and indeed wider kinsfolk and society at large. In such circumstances there is no doubt who the mother is and who the father is; nor is there serious doubt about the roles expected of father, mother, son, daughter, husband, wife, brother and sister. This is not to claim that these expectations are always met but most would find little difficulty in providing a general description of what behaviour would normally be present in each role. Of one thing we are certain, underlying the behaviour and values described will be the assumption that the mother combines the genetic, carrying and nurturing functions of motherhood; that the father combines the genetic and nurturing functions on the male side; that the two are married; that they enjoy an exclusive sexual relationship; that the children of the marriage are 'theirs' and that the brothers and sisters are also the product of that exclusive sexual relationship. What is interesting about this list is that all items would normally have been assumed and few, if any, would have been made explicit. It is the arrival of artificial reproduction techniques which operate outside the institution of marriage which has exposed these assumptions about family relationships. To be sure, societal norms are sometimes broken, and children may arise from adulterous unions, but it is usually accepted that such a disregard of the norms is likely to cause disruption of family life, and disregard of the norms is not encouraged.

EFFECTS OF ARTIFICIAL REPRODUCTION ON THE FAMILY

We noted in the last chapter that the separation of reproduction from sexual intercourse whereby the one is not necessarily the result of the other has resulted in a strange situation for, depending on the definition used, a child produced by artificial means can have anything up to five parents. If the genetic, carrying and nurturing mother and the genetic and nurturing father can all be different people the question may be reasonably asked: who is the mother and who is the father? That this is not a rhetorical question is clear when we recall that the person so identified is subject to a range of obligations and possesses certain rights, and that these can only be given up or ignored in exceptional circumstances. If more than one kind of mother exists, which shall have precedence? Will it be the genetic mother or the nurturing mother? Does the carrying mother who is providing the first social, as opposed to the genetic, link with the developing child have any rights and obligations? The carrying mother may be performing an essential biological function but the formation of the mother–child link is not solely a biological one. The attitude of the mother toward the baby she is carrying plays an important part in the baby's development, and such an attitude is reflected in behaviour relating to her own body. Many pregnant women avoid nicotine, alcohol, analgesics and other drugs in order to provide as healthy an environment for the growing baby as is possible. Less obvious, but equally important, is the need to avoid stress and the desirability of having a positive and welcoming expectation of the new baby's birth. The maternal–child bond can perhaps be over-emphasised but generally it is accepted that the period from conception to birth is vitally important for the future development of the child. It is also generally recognised that the total separation of the purely biological from the psychological and social is not possible, and therefore the very early development of the child during pregnancy should not be dismissed as being of no subsequent social consequence to the child.

This early relationship in the child's development is a two-way process, for not only is the mother forming a relationship with the child while she carries it in her own body but also the child is forming a relationship with the mother. It is perhaps not inappropriate to point out that a child/father relationship may also have a deeper significance if the father who is subsequently to nurture the child after its birth is present and is, in a more limited way, a participant during the pregnancy. This suggests that the mother who carries the child during pregnancy with the declared aim of giving the child to another woman at its birth, occupies a position which is particularly difficult to justify in terms of the parent/child relationship.

Unlike the mother role the father role encompasses just two functions which are therefore easier to interpret. There is no comparative relationship between the father and his child as exists for the mother during pregnancy. Perhaps it is for this reason that so much emphasis is placed on the genetic implications of fatherhood.

In the usual situation the man producing the semen which provides half the potential of the new conception is also the man who is to nurture the child once it is born. If the child is not created out of the man's semen then the claim that he is the 'father' of the child is difficult to justify. To be sure, there are exceptions but where these occur special arrangements have usually been made so that the man can publicly declare his wish to take full paternal responsibility for the child. This practice has been known for centuries and is what we call 'adoption'. The relationship between a parent, whether

51

as father or mother, and an adopted child is not the same as that where a genetic link exists but its strength is not evenly balanced; for whilst frequent reference to an adopted son or an adopted daughter is made, the appellation of adoptive father or adoptive mother is less common. It appears that the direct causal genetic link from parent to child is exerting an influence even when, in a particular case, it is absent. The term 'adopted' son or daughter is a negative indication that the child is not genetically linked to the parent. Once a public acceptance of the adoptive relationship has been made, the legal rights of the child in relation to its adoptive parents are the same as for a child conceived by the parents. But reference to legal rights is not the same as an understanding of social organisation. It is for social and psychological reasons that the adopted child has rights relating to the identification of his genetic parents. If the genetic parentage were not considered important it is hard to understand why this particular 'right' has received so much emphasis in recent years.

Having emphasised the role of providing the genes and of carrying the child during pregnancy as being necessary to the understanding of parenthood, we come to the most important role of all; that of nurturing the child after its birth. Both the male and the female have a role to play and the ways in which these roles are defined by society are well known and can be emulated. Parents are expected neither to neglect their children nor to indulge them over much nor punish them too severely; in short they are expected to conduct themselves with care, consideration and moderation, with insight and amiability, being firm when firmness is required and supportive and encouraging at all times. To be a parent requires a special relationship with the child, but if the genetic and/or carrying component of the child's development before its birth are not also associated with the same nurturing parent, then doubt is often expressed about the child's real parentage. It is surprising how pervasive this recognition is; many infertile husbands who have become a parent through AID frequently refer to the donor as the 'real' father. It seems that what we consider as being the most important role of parenthood, namely the nurturing of the child, is insufficient of itself to endow full parent status. The nurturing role may be seen *socially* as the most important but the genetic role is also an essential component, and this is recognised by both parent and child.

Most family members recognise their relationship with each other in two dimensions, past and present relationships, and both are associated with perceived genetic links. In answer to the persistent question 'Who am I?', most would eventually reply by reference to their position in familial terms. 'My family came from . . . where my father was a . . . and my mother . . .' etc. Each of us has a past in terms of family history and a present in terms of our current family relationships. These two aspects of family life are taken so much for granted that they are often left as unspoken assumptions which underlie the subjective feelings each of us possesses concerning our own particular family 'tree'.

It seems that in the normal family situation the mixture of genetic and nurturing functions is difficult to separate when viewed from a social perspective. But recent technological advances have enabled a division of these functions in ways that were not previously considered except in the science fiction novel. The effect upon the social understanding of family life has been profound not only because of the division of functions previously combined but also because an element of deception has been introduced, the social consequences of which have been little studied.

We have said that the relationships in the family are both important and special ones. The task of socialising children, and bringing them up to display behaviour that is both conforming and socially acceptable, is the prime task, not only of immediate family members, but of the wider family of kinsfolk. This is so partly because the law prescribes it but also partly because custom and practice decree it. The protection and care of young children is, to say the least, one of the most important tasks of society. In the process children learn they have duties and obligations towards other members of the family and through them to other members of society. They also are aware of what they may expect from family and non-family members. For the well-being of the child the familial environment should be stable and predictable. In short, the relationships are essentially trusting ones. This does not mean that members of the family have no privacy but that the relationships are open, spontaneous and not essentially calculating with a personal advantage in mind.

We may say that the well-being of society depends upon families bringing up children in such a manner that in character and behaviour they are capable of making a contribution to their fellow men in the wider sphere of relationships at work and elsewhere. For this to happen the trust that lies at the heart of family life must be preserved. The relationships within the family are complex, more so than is commonly supposed. Yet what is important in all families is the understanding, or assumption, that a particular relationship exists by right. Even in times of stress, perhaps especially at such times, this is not lost sight of but recalled and may be recognised by those who are not members of the family. We may conclude that if procedures are introduced which undermine these assumptions about relationships the effects on family life may be profound. What keeps the family group together is the tacit, unspoken assumption of a special relationship. The basis of this special relationship has much to do with the subjective interpretation of what creates family life. As we have indicated, the blood relationship as well as the rules which support the group, some of which are formally supported in law, are at the root of the matter. If either of these bases is eroded the implications for the family members may be serious.

Part II

THE
SOCIAL IMPLICATIONS OF AID:
A RESEARCH PROJECT

4

Social Profile and Experience of Couples Undergoing AID 1940–1980

The lack of evidence about the long-term outcome of AID and its effects on the families and children so produced, was stressed by all the committees which inquired into AID. Speaking in the House of Lords debate on AID and Legitimacy in 1949 the then Archbishop of Canterbury said, 'My commission observed, very truly, that as yet there is little evidence available upon which to judge the sociological and psychological effects of AID. There are too few observed cases, and too few cases observed for a sufficient length of time.' The Feversham Report of 1960 stated, 'there is no doubt that all discussion of the practice is at present greatly handicapped by lack of information about what has subsequently happened to the families of those women who have received AID. . .' The Peel Report of 1973 commented:

The Panel has had evidence from several sources urging that, for psychological reasons, there should be no follow-up of children born of AID. It understands but does not entirely agree with these views. Information must be obtained on the genetic effects especially where frozen semen has been used and it is important to learn the effects, in human terms, on the development of personal relationships in families resulting from the use of AID.

One practitioner who first began to provide an AID service in September 1940 continued to provide such a service until March 1982. Her work is unusual in this country in that right from the outset she attempted to keep in regular contact with couples who had a baby following AID in her practice. Couples were encouraged to write at least once a year and many couples brought their babies and growing children to visit this gynaecologist when they were on holiday in the area. She received regular information about the development of the child and the well-being of the family concerned. There was, therefore, a great deal of retrospective information available and an atmosphere of mutual goodwill between the practitioner and these families.

Towards the end of the 1970s this practitioner approached the authors (with whom she has a long history of co-operation in research endeavours) and suggested that this social data, carefully collected over a lifetime of work, should be systematically analysed. Other practitioners had analysed and reported on their own work, mainly assessing success in terms of pregnancy rates, but there had not previously been an independent investigation of the social issues surrounding AID.

There was one further factor which was causing concern to this gynaecologist, now nearing retirement after a pioneering career devoted to helping women have the number of children they wished to have, neither too many nor too few. Certain unusual cases of AID were receiving publicity in the press; single women who did not wish to marry, and women in lesbian relationships, were having children by AID. Cases of surrogate motherhood, where artificial insemination techniques had been used to achieve a conception, were also receiving publicity. However, the vast majority of cases of AID still take place within a marriage relationship with a husband and wife bringing up the AID child or children in a traditional family setting, the husband in his role of nurturing father taking on all the attributes and responsibilities of fatherhood. There was a fear that access to AID for these married couples might be regulated according to the needs presented by the more unusual 'fringe' cases of AID rather than according to the needs of the vast majority of couples who used the service. It was therefore important to know something of the family experience of these couples.

An attempt was made to provide a social profile of the people who had attended for AID at this one practice over a period of forty years, and to determine the characteristics of this population in terms of how successful, or unsuccessful, their use of AID had been. In the past, couples have usually kept their involvement in AID a secret and relatively few have been open (with either family or friends) about their use of the procedure. An attempt was made to examine the reasons which lay behind this perceived need for secrecy, the extent of the secrecy, the means by which the secret was kept and how secrecy affected the life of the family members. The social implications of AID as they affected the couple and their child, and their relationships with wider family and friends were examined from the perspective of the *parents* of AID children. Because of the secrecy which surrounds AID it was not possible to study directly the implications of AID as they are perceived by the wider family, or by the child itself. However, during the course of the study, contact was made with three young adults who were aware that they had been conceived as a result of AID, and who were willing to discuss their experience. By assessing the social significance of AID, a greater awareness of the needs of a significant proportion of childless couples and their families would be facilitated; this may in turn lead to improvements in the infertility services provided for these couples.

METHODOLOGY

The study was undertaken in two parts, the first concerned with a profile of AID couples who had attended this single Devon practice for advice or treatment connected with infertility between September 1940 and December 1980. Background information from these records was collected to see if a detailed study was feasible. The second part of the study consisted of interviews with a sample of couples who had successfully used AID. This chapter describes the first part of the study only, and Chapters 5, 6 and 7 provide an analysis of the information collected in the interviews.

It proved possible to collect consistent retrospective data relating to demographic and social background factors, and to describe the changing pattern of AID provision over time. This data collection was undertaken by the member of the research team who was already working with the practitioner. A system was devised which permitted analysis of available information whilst at the same time maintaining the anonymity of the couples concerned.

The records of a total of 986 couples who had received AID during the four decades 1940–80 were examined; 87 of these couples had concurrently been treated with artificial insemination using the husband's semen also (AIH) and they were removed from the sample leaving 899 couples who had received AID only. Details of the age of the couple, duration of the marriage, any previous marriage, any previous children, reason for AID referral, area of residence, and occupation of husband were collected. The date of the first insemination, total number of inseminations, reason for stopping inseminations, and the time between first insemination and stopping were also recorded. Such a sequence of inseminations, over a period of time, ending in either a conception or the discontinuation of the procedure for any other reason, was defined as an AID 'session'. Because the long distances travelled by most patients ruled out the possibility of repeat visits, usually only one insemination was performed during each menstrual cycle. During some cycles no insemination was carried out because it was not possible for the patient to attend; the usual reasons for this were holidays, snow-blocked roads or illness. Most AID sessions therefore cover a slightly longer period of time (in months) than the number of inseminations might suggest.

It was the normal practice to use fresh semen but frozen semen was used when suitable fresh semen was not available at the appropriate time. Another study (Jackson and Richardson, 1977) has compared the results obtained using fresh and frozen semen but no differentiation between the two is made in the present study.

The case-records also contained a considerable amount of correspondence relating to the couple's experience of AID, the subsequent development of their child/children, and the progress of their family life. This correspondence provided valuable insights into those issues which were thought to be important by the couples concerned.

RESULTS

The retrospective data relating to 899 couples who had received AID between September 1940 and December 1980 were analysed to provide a social profile of these couples and the outcome of their AID experience. The outcome of AID in these 899 patients is shown in Table 4.1. For those patients who conceived, Table 4.2 shows the outcome of pregnancy. A total of 480 live children were born; 240 couples had one child only, 101 couples gave birth to a second child, 11 couples had three, and 1 couple had five children. (The firstborn were twins who died when a few days old). Of the

Table 4.1   *Outcome of AID*

Session	Pregnancy		Gave up		Lost to follow-up		Still continuing	
	no.	*%*	*no.*	*%*	*no.*	*%*	*no.*	*%*
I N = 899	391	43·5	399	44·4	68	7·6	41	4·6
II N = 299	137	59·8	64	27·9	9	3·9	19	8·3
III N = 47	33	70·2	8	17·0	1	2·1	5	10·6
IV N = 9	5		2		2		—	
V N = 1	1		—		—		—	

480 babies born alive, six suffered from congenital abnor-malities and six from abnormalities caused by birth trauma. The majority of these infants died but four lived and were cared for by their families. Eight of these parents returned for further AID; four couples subsequently had a healthy baby but the remainder gave up without achieving a preg-nancy. Following their first AID session seven couples gave birth to a still-born infant; six of these couples came back to try a second time and four subsequently gave birth to a live baby.

Sessions varied in the number of inseminations from one to ninety-one; for the first session the average number of inseminations was nine and the mode two. Most conceptions occurred quickly, over half (51·8 per cent) within five insemi-nations or less and more than three-quarters (76·7 per cent) within ten inseminations or less; nevertheless there were four pregnancies among the thirty couples who persisted for more than thirty inseminations. One third of the total number of couples went on to a second session of AID and 93 per cent of these couples had conceived during their first session. The likelihood of achieving a pregnancy increased among couples who returned for second and third sessions

Table 4.2 Outcome of Pregnancy*

Session	Live birth (female)		Live birth (male)		Twins		Miscarriage		Still birth		Pregnancy continuing
	no.	%	no.	%	no.	%	no.	%	no.	%	
I N = 391	163	42·8	167	43·8	2	0·5	43	11·3	6	1·6	10
II N = 137	61	46·6	52	39·7	1	0·8	16	12·2	1	0·8	6
III N = 33	12	42·9	14	50·0	—		2	7·1	—		5
IV N = 5	4		—		—		—		—		1
V N = 1	—		1		—		—		—		—

* Percentages are of completed pregnancies.

of AID (Table 4.1). This may be due to wives who were themselves sub-fertile being screened out during the first session, but it is possible that psychological stress also played a part. The topic of stress is discussed more fully in Chapter 7, but it has been shown that psychological stress can delay or even prevent ovulation. Wives who returned for further AID after successfully becoming pregnant the first time would presumably be less anxious and stressed, and therefore more likely to conceive. This hypothesis is supported by the evidence that wives who gave birth to a live baby after their first session of AID were more likely to conceive during their second session than wives who had had a miscarriage after the first session (67·2 per cent compared with 45·9 per cent). Once conception had occurred, wives who had previously miscarried were just as likely to have a live birth as women who had previously had a live birth (88·2 per cent compared with 86·2 per cent).

DISCONTINUATION

Most couples who gave up did so quickly (38·7 per cent within five or less inseminations, 60·4 per cent within ten or less inseminations). This suggests that these couples actively changed their minds, rather than gave up because the procedure was unsuccessful. Table 4.3 shows the reasons offered for giving up in Session I; it may well be that there were deeper and more complex reasons which were not explored. A sizeable proportion of couples gave up to adopt; some of these had merely been accepted by an adoption agency and placed on a waiting list, but others had been offered a baby. A substantial number of couples gave up because they found the nervous tension and discouragement of hoping but failing to conceive each month more than they could bear. As one wife remarked, 'I don't want to spend Christmas thinking "Am I, am I [pregnant]?" It's a nightmare the whole time you are coming down here; you're waiting for your appoint-

Table 4.3   *Reasons for Giving Up AID–Session I (N = 899)**

	No. of couples	% of total sample
To adopt	118	13·1
Travelling distance too great	80	8·9
Discouragement and nervous tension	69	7·7
Financial difficulties	19	2·1
Husband now disapproves	3	0·3
Marital disharmony	3	0·3
On medical advice	78	8·7
No explicit reason given	46	5·1
Conceived naturally	8	0·9
Other reason	32	3·6

* NB. Some couples recorded more than one reason.

ment, and then you're waiting to see if you are pregnant.'
Very few couples gave up because of financial pressures; the
fees charged in this practice were low and the bulk of the
costs incurred were travelling expenses. It was rare for
couples to give up AID because of marital disharmony or
disagreement. Other reasons for giving up included cases
where the husband went to sea or received an army posting
abroad and the wife felt it would be imprudent for her to
conceive during his absence. Some couples found it difficult
to get time off work and others felt they had to keep the visits
secret and found it too difficult to do this. One husband
confessed he felt uneasy at deceiving his parents, and
another explained that his wife 'found it too difficult to
accept the unknown'.

During their second session of AID, four couples gave up
because of marital disharmony. This is only a very small
proportion, but greater than in Session I. Among these four
couples, two wives had given birth to still-born infants after
their first session of AID (one with multiple deformities), and

one had had a termination of pregnancy because a spina-bifida abnormality had been diagnosed in the foetus. The birth of an abnormal child puts great strain on any marriage, but it seems likely that the birth of an abnormal child following AID (of which there is the normal risk) would cause even greater stress. A couple who already had two AID children gave up their third session of AID because of marital dishar-mony. A return for further AID is often used as an indication of 'success'; this may be so in many cases but it does not necessarily follow.

The proportions of couples giving up because of the dis-tances and travel involved were similar for the first and second sessions of AID. This is surprising as couples coming for a second session would already be aware of this incon-venience. However the majority of these couples now also had the needs of a small child to consider, and this may have caused additional, unforeseen problems. Many couples find baby-sitting arrangements difficult to make as they feel obliged to keep the real purpose of their journey secret. Wife 664 said, 'Although I could leave him [AID son] with grand-parents, it's difficult to explain where I'm going – you run out of reasons after a while. It's a big problem now. Some-times you have a late appointment and by the time you get back it's past [the time] when all the shops have closed and parents say "Oh, where have you been?"'

SOCIAL PROFILE OF SAMPLE

RESIDENCE

Only 28·7 per cent of the couples lived in Devon; a further 34·4 per cent came from the South-West region and the remainder from all parts of the United Kingdom. Five couples travelled from overseas. The likelihood of achieving a pregnancy decreased slightly as the distance which couples lived from the clinic increased (Table 4.4). Ledward *et al.*'s

Table 4.4   *Effect of Area of Residence on AID Experience*
(Session 1)

Area of residence	% of total sample attending	% couples achieving pregnancy (completed sessions only)	% couples attending for 10+ inseminations
Devon	28·7	47·5	37·2
S. W. England	34·4	48·8	37·5
S. England and Wales	21·5	44·3	27·5
London	4·1	37·8	21·6
Midlands and E. Anglia	5·5	35·4	24·4
N. England	3·3	43·3	10·0
Other UK	1·2	36·4	—
Overseas	0·6	—	—
Not stated	0·8	—	—

(1982) study of couples attending the NHS AID clinic at Nottingham also produced a similar finding. Couples living nearer to the practice were also more likely to persevere with AID visits; 37·2 per cent of Devon couples persisted for more than ten inseminations, but this figure decreased as distance from the practice increased (Table 4.4). The provision of AID centres remains limited, and many couples still have to travel long distances to be seen. A survey in 1977 showed that Wales, the North-East, East Anglia and Southern England still had no AID service. During the decade 1971–80 64·1 per cent of the couples attending this Devon practice still came from outside the county (Table 4.4).

SOCIAL CLASS

The couples were classified according to the husband's occupation using the Registrar General's classification.

Non-manual workers were over-represented and comprised
52·4 per cent of couples; skilled workers formed by far the
largest proportion of the manual working couples (Table
4.5). In the absence of any research evidence one can only
speculate why working-class couples, especially those who
are not skilled, are less likely to take up the option of AID. It
may be that manual working couples are hesitant to become
involved in a controversial procedure. Working-class
couples who do come for AID do not begin later in their
marriage than middle-class couples (Table 4.5) so it is unlikely that working-class couples are just unaware of the
possibility of AID. It may simply be that they cannot afford
it. Fees were very low in this practice, nevertheless manual
workers (especially the unskilled) were more likely to give
up for financial reasons and unemployed couples were most
likely to do so. Professional couples were most likely to have
travelled to this practice from outside the South-West
region, and unskilled or unemployed couples were least
likely to have done so. Not surprisingly these groups respectively were also the most and least likely to discontinue AID
because the travelling distance was too great. It is interesting
to note also that middle-class couples were more likely to
give up AID in order to adopt (Table 4.5). This may be
because middle-class couples look more favourably on the
alternative of adoption than do working-class couples, or it
may be that adoption selection procedures favour middle-
class couples.

AGE

The husbands in the survey tended to be considerably older
than the wives; 58·8 per cent of husbands were 30 years of
age or over compared with 30·4 per cent of wives; 116
husbands were ten or more years older than their wife
and 15 were more than twenty years older. The likelihood of
achieving a conception decreases with the increasing age of

Table 4.5   Effect of Social Class on AID Experience (Session 1)

	% of total sample	% couples achieving pregnancy	% pregnancies ending in miscarriage	% couples married 5+ yrs at 1st insemination	% couples discontinuing for financial reasons	% couples travelling from outside SW region	% couples discontinuing due to distance	% couples discontinuing to adopt	% couples discontinuing on medical advice
Professional N = 94	10·5	43·6	14·6	39·4	0·0	54·3	13·8	14·9	5·3
Semi-professional N = 228	25·4	43·9	8·2	43·4	1·3	33·7	10·5	15·4	7·0
Skilled non-manual N = 148	16·5	45·3	10·6	37·2	1·4	34·5	7·4	16·9	9·5
Skilled manual N = 315	35·0	40·3	10·4	36·5	2·9	37·5	7·9	10·5	8·6
Semi-skilled manual N = 73	8·1	47·9	9·4	47·9	2·7	38·4	8·2	9·6	9·6
Unskilled manual N = 31	3·4	48·4	33·3	29·0	6·5	19·4	0·0	12·9	16·1
Unemployed N = 6	0·7	50·0	0·0	0·0	16·7	0·0	0·0	0·0	0·0
(Not stated N = 4)	0·4								

Table 4.6 *Effect of Age on AID Experience*

Age group of wife		% of total sample	% couples achieving a pregnancy	% couples discontinuing	% couples discontinuing due to nervous tension	% couples discontinuing to adopt
19 years or under	N = 5	0·6	20·0	80·0	20·0	0·0
20–4 years	N = 204	22·7	48·0	39·7	6·4	9·8
25–9 years	N = 417	46·4	47·5	39·3	6·7	12·5
30–4 years	N = 202	22·5	38·1	50·5	7·4	14·9
35–9 years	N = 60	6·7	23·3	66·7	16·7	25·0
40 years and over	N = 10	1·1	10·0	90·0	20·0	10·0
Not stated	N = 1	0·1				

the wife (Table 4.6). The pregnancy rate fell after the age of 30 years and particularly so for those over 35 years; wives in their 20s were more than twice as likely to conceive as wives in the 35–9-year age bracket. Not surprisingly wives in the 35–9-year age group were more than twice as likely as younger women to give up AID because of discouragement and nervous tension. If couples are to have the best chance of a pregnancy it is important that information about AID should be readily available so that couples can be referred while the wife is still relatively young. Many couples experience considerable difficulty in obtaining information about AID and many doctors are hesitant (or unwilling) to initiate the suggestion. As one husband remarked, 'I've said several times, if only we'd known of the existence of such people ten years ago. That's the only thing I've got against it – we should have known years ago.' Couples who have heard about AID and wish to embark upon the procedure may still experience considerable delay; the scarcity of AID clinics, particularly within the NHS, means that waiting lists may be long and in some cases couples who do not live within a certain catchment area do not qualify for entry on to a particular waiting list.

MARRIAGE

All couples were in what was considered to be a stable marriage. Almost 40 per cent of couples had been married for more than five years when they were referred for AID. Most couples (89·5 per cent) were married for the first time but in 10·5 per cent of couples, one or both partners had remarried. It was three times as likely for husbands to be in a second marriage as wives; seventy-seven men were remarried compared with twenty-seven wives. Twenty-five men had had a vasectomy following the birth of the children of their first marriage, but about one-half of the men who had remarried had been found to be infertile during their first marriage. It may be that childless marriages are more easily and readily dissolved but it is also likely that the stress of male infertility and involuntary childlessness increases the risk of marriage breakdown.

The proportion of couples who experienced marriage breakdown after AID is not known precisely as contact was not maintained with all couples but it is thought that the incidence among AID couples is lower than that found in the general population. The reason for this is not clear; it may be that couples who opt for AID are particularly committed to each other and are determined to surmount their problems of childlessness together. In this study several husbands had discussed divorce with their wives when it became clear that they could not give them children. It may be that some infertile couples who are less committed to each other take the option of divorce in preference to AID. For these reasons a lower rate of divorce among AID couples could be expected but the reasons given must remain speculative at the present time.

CHANGES OCCURRING OVER TIME

The number of couples beginning AID at this one practice almost doubled each decade (Table 4.7). The majority of couples (78·4 per cent) were referred for AID because the husband was found to be azoospermic. This proportion has remained fairly constant though it fell to 71·3 per cent between 1976–80. No couples were referred because of genetic factors until the late 1950s. The numbers, while remaining small, have steadily increased since then, presumably as a result of increased provision of genetic counselling facilities. Couples were first referred following vasectomy in the early 1970s and by the end of the decade accounted for almost one in ten couples (9·2 per cent). The proportion of couples travelling to the practice from outside Devon increased in the 1950s and 1960s, presumably as this

Table 4.7 Social Profile Over Time

Date of commencing AID	No. couples commencing AID	% couples coming from outside Devon	% couples husband manual worker	% couples married 5+ years	% couples married 10+ years	% wives aged 30 yrs and over	% couples giving up to adopt	% couples achieving a pregnancy
1940–50	76	67·1	42·1	43·4	17·1	53·9	25·0	43·4
1951–60	128	83·6	39·8	46·1	12·5	46·9	17·2	49·2
1961–70	256	78·7	46·5	41·5	8·9	30·2	17·0	44·6
1971–80	437	64·1	49·4	36·0	5·3	21·3	7·5	41·0

practitioner became more well known. The proportion decreased during the 1970s, probably as a result of increasing provision in other areas, but still remains high at 64·1 per cent of couples. Whilst non-manual couples are still over-represented in the sample, the proportion of manual couples availing themselves of AID has gradually increased. The proportion of older women beginning AID has steadily decreased over the years. In the 1940s just over half the wives were aged 30 years or over when they commenced AID but by the 1970s the proportion had decreased to just over one-fifth. The proportion of couples who had been married for more than five years when they commenced AID remained fairly steady until the 1970s when it decreased slightly. The proportion of couples married for more than ten years when they commenced AID has steadily decreased over the years. The proportion of couples giving up AID in order to adopt declined markedly during the 1970s when there were fewer babies available for adoption (Table 4.7).

The changes occurring during these four decades would suggest that recourse to AID by infertile couples is becoming increasingly common. The practice is no longer restricted to infertile couples as AID is being used to acquire children in second marriages following vasectomy. The take-up of AID by working-class couples is increasing but this is still mainly restricted to skilled workers. The proportion of younger wives beginning AID has steadily increased and it would seem, therefore, that information about this possible solution to childlessness is now somewhat more readily available. However, the restricted provision of AID means that couples still have to travel long distances in order to obtain the service they seek.

5

The AID Family

An examination of the retrospective data contained in case-notes (described in Chapter 4) made possible the construction of a generalised social profile of the type of people who had attended for AID at this one practice over a period of forty years. It was also possible, in a generalised way, to describe the AID experience of these people and to define trends relating to characteristics which made a successful use of the service more or less likely. A further objective of the research project was to learn more about the AID experience of individual families and to examine this experience at greater depth. How do couples come to terms with the idea of having their children by AID? How do they cope during the period of time spent attending for inseminations? Having succeeded in becoming pregnant how does the fact of AID affect the family life of the couple and their child? How does it affect their relationships with their wider family and with friends? And, perhaps most importantly of all, how does it affect the life of the resulting child?

Because couples usually keep their involvement with AID secret, and most AID children are not aware of their AID origins, it was not possible to discover the answer to this last and most important question by talking directly to AID children themselves. Nor was it possible to discover directly what was the view of close family relatives and friends. It was only practicable to attempt to interview AID couples themselves. The social implications of AID as they affected the couple and their child, and their relationships with wider family and friends would, therefore, have to be interpreted from the perspective of the AID couple.

The secrecy surrounding AID determined the design of the research strategy in several other respects. It was important that the parents' wish for secrecy was respected, and vital that no children were accidentally made aware of their AID origins because of the conduct of the study. This meant that couples who had children old enough to be made suspicious by an approach to their parents could not be interviewed. The only exceptions to this were couples who had AID children over the age of 18 years who had 'flown the nest' and who were now independent of their parents. Consideration was given to the idea of investigating the experience of the parents of school-age children by means of a postal questionnaire, but it was feared that some questionnaires might inadvertently be opened or discovered by relatives or even by the children themselves. No matter how carefully worded the questions were, it was conceivable that such an event might lead to accidental exposure of the couple's secret, and to the unintended disruption of their family life.

INTERVIEW SAMPLES

The group of older couples with grown-up children available for interview were to some extent self selected as they had voluntarily and conscientiously maintained contact with the practitioner over many years. This could mean that their experience was biased and not representative of the experience of all couples. To counter this bias all couples attending this practice who had young, pre-school children born in the previous four years, whether or not they had maintained contact with the practitioner, were identified. This meant that all couples who had a baby (following donor insemination in this practice) during the period between 1 January 1977 and 31 December 1980 were listed. A letter was sent by the practitioner asking these ninety-two couples if they would be willing to discuss, in confidence, how they felt about their AID experiences. This letter was carefully worded and made no explicit mention of AID in case it was inadvertently seen by a relative or friend. A second group of nineteen couples, who had continued to keep in touch with the practitioner and who were the parents of AID children all now over the age of 18 years, were also identified and sent a request for interview.

INTERVIEW TECHNIQUE

It was important that the interview should give the couple an adequate opportunity to express all their views about AID and to express them freely. A tightly structured interview was therefore inappropriate. Nevertheless the discussion needed to be focused on the subject of AID, and so a list of topics to be explored was drawn up. These topics were not approached formally or in any particular order. The couples were given the opportunity spontaneously to raise issues, and these were explored and followed up. The list of topics was used merely to prompt new areas of discussion and to ensure that all areas were covered in each interview.

Couples who expressed willingness to be interviewed were telephoned in order to make an appointment, to establish an initial contact, to give a brief explanation of the reasons for the study and to provide reassurance about confidentiality of the interview. Some wives were pregnant again following further AID and these couples were not interviewed until after the birth of the new baby. The authors were responsible for conducting all the interviews which mostly took place in the couple's own home. Couples were interviewed together by two interviewers, one male and one female. It was considered inappropriate to undertake separate interviews of the husband and wife because of the uncertainty this might create between them. Many couples lived a considerable distance from the practice, and interviews took place (mainly at weekends or in the evenings when the husband was available) throughout the South-West, South Midlands and South-East England. A few couples preferred not to be interviewed at home. These were mostly couples who were travelling to the practice for further AID in order to have another baby. These couples were interviewed at the home of one of the interviewers as it was felt that an informal, domestic environment would help to keep the interview more relaxed and informal.

The topics covered in the interview included:

Initial knowledge of AID provision
Experience of infertility diagnosis
Decision-making about AID involvement
Adequacy of counselling
Initial anxieties about AID
Consideration of adoption and comparison with AID
Family structure – proximity of relatives (geographical and social)
Degree of secrecy maintained and reasons for this
Problems encountered in maintaining secrecy
Extent and frequency of discussion about AID involvement
Advantages/disadvantages in telling close family and friends
Intention to tell, or not to tell, the child and reasons for this decision
Legal status of AID children
Code of practice for AID provision

Interviews usually lasted for one-and-a-half to two hours and were tape-recorded. It was considered that note-taking during the interview would hinder the conduct of the interview and might distract or inhibit the couples in the free expression of their views. Notes made after the interview, even if written up immediately, would inevitably have been incomplete. The tape-recorder, once switched on, could be ignored and was usually quickly forgotten. Great care was taken to ensure that confidentiality was maintained. Christian names only were used in the discussion; one interviewer only, for administrative reasons, knew the full names of the couples. The interviews were transcribed by this interviewer without secretarial help, as it was felt that mention of place-names, occupations, unusual family events or illnesses could, together, accidentally reveal the identity of a couple. All documents and tapes were kept locked away.

Tandem interviewing of a couple worked well. The great majority of interviews were relaxed, informal and informative. It is likely that by interviewing the couples in their own homes, on their own ground, the couples were given a greater sense of security and of being in control of the situation. Interviewing the couple in their own home had the added advantage of allowing the interviewers to observe the 'atmosphere' of the home, and to view at first hand the interaction of some of these parents with their young AID children.

Many couples found the interview helpful and therapeutic as it had provided them with their first opportunity to talk out their experience at length. In a few cases the husband displayed some hostility early in the interview. This was signalled either by a degree of aggression in the response to questions, or by an initial reluctance to enter into discussion; but in all cases this eased as the interview progressed. In a few cases the couple, whilst being superficially co-operative and obliging, appeared to be unwilling (or unable) to explore and expose their feelings fully. It would have been preferable if follow-up interviews could have been arranged in order to explore certain issues raised in discussion more deeply, but in the present climate of opinion this was felt to be impracticable.

THE EXPERIENCE OF PARENTS OF AID CHILDREN

Of the 92 couples who had given birth to an AID child between January 1977 and December 1980, 81 were successfully contacted. Of the remainder, 8 had moved away and the new address was not known, and 3 did not reply. Eighteen of the 81 couples who were successfully contacted refused to be interviewed. Twelve couples gave no reason for their refusal but the others gave one or more reasons. Most felt this was an exceedingly private area of their experience which they wished to keep confidential. Others stated they wished to forget their experience and not to be reminded that their children were conceived by AID. Four couples stressed that despite their refusal, they were very happy with their children and all was going well. The proportion of successfully-contacted couples who refused to be interviewed was small; 78 per cent of couples were willing to be interviewed. It is therefore likely that the results obtained from the interviews are representative of the experience of other couples who attended this particular practice for AID.

Sixty-three couples were willing to be interviewed and 57 interviews have been completed. Of the remaining 6 couples, 3 had moved away too far to make an interview feasible, 1 couple achieved another pregnancy after further AID and by the time this baby was born had changed their minds about interview, 1 couple live in a gypsy community and will contact the research team when they travel to the South-West, and one couple were unwilling to be seen during the day when their small child was up and about, but lived too far away for an evening visit to be feasible. However, this couple wrote a long and detailed account of their AID experience. Among the 57 couples interviewed, 35 couples had one AID child, 20 couples had two children, 1 couple had three and 1 four children at the time of interview. (Some of these children were born after December 1980, therefore the numbers do not correspond exactly with the figures given in Chapter 4). Two couples also had an adopted child, 2 couples a foster child, and 1 couple two step-children of the husband's first marriage living with them. The majority of couples (53) were referred for AID because the husband had been discovered to be infertile. Three couples were referred by a genetic counsellor; 2 of these couples had already given birth to a handicapped child. One couple sought AID because the husband had a vasectomy during his first marriage.

The overall impression gained from the interviews was that the vast majority of couples were immensely grateful that their involuntary childless state had been resolved. Almost without exception they had no regrets about their decision to employ AID, and with hindsight would make the same decision again. This did not mean, however, that they had discovered no hitches or problems. Many couples felt that some of their experiences, and the way they had dealt with certain situations, had not been ideal; given the opportunity, they would have wished to change things in certain respects. All the husbands (and wives) appeared to have accepted the children willingly and happily; indeed some of the fathers had a particularly close relationship with their children and appeared to be deeply involved in child care and family life. Because their children had been achieved after considerable heartache, and after much effort, they were particularly valued and loved and the couples tended to find parenting particularly rewarding and satisfying.

In the sample of couples interviewed, non-manual workers were again over-represented; 30 husbands were non-manual and 25 husbands manual workers. Two husbands were unemployed because of physical disability. The majority of manual working husbands were skilled workers and most couples were buying their own houses; only 9 of the 57 couples interviewed lived in local authority housing. This objective finding confirms the subjective impression of the interviewers that the majority of manual working couples who were interviewed had a life-style which, to a great extent, followed middle-class values. This analysis of the interview sample confirms the finding of the retrospective survey that couples who avail themselves of AID tend to be middle class.

THE DECISION TO ACCEPT AID

Most couples stated that information about AID was difficult to find. When seeking medical advice about infertility only one-half of the sample (29 couples) had the option of AID suggested to them by their medical advisers. The other couples heard about AID from the media, usually newspaper or women's magazine articles or television documentaries, and they themselves made the initial approach to their doctors. Four of these couples reported a hostile reaction. One wife said, 'My doctor tended to put me off. He was far more keen on us going for adoption. I had to fight. I had to be very adamant and say "I want to go to her. Will you please refer me?" – and he did.' Many couples searched for information about AID but found it difficult to obtain. One wife, who had been to several libraries in a vain search, expressed the position well. 'There were just statistics, one or two books on figures – 10 per cent of 10 per cent – that sort of thing. There was nothing of an emotional nature that one could seek comfort from or help from. Lots of books on infertility which we read, and hoped there might be help for us but of course our case wasn't even possible.' The main sources of information were short newspaper articles or TV documentaries; the couples clutched at these straws but found them inadequate. One husband, a professional man, said he was

aware that there was such a thing as AID from magazine
articles . . . 'but I can't recollect ever having seen how you
get in touch with how it all starts. It's something for other
people if you aren't very careful.'

Because information was restricted, mainly to factual
matters, it was difficult for couples to make a fully considered decision. When couples were referred to this practice
for AID, both spouses together had a lengthy initial interview
with the practitioner at which they were able to ask questions. But by this stage they had more or less decided to go
ahead with AID and their questions tended to be centred on
the medical and factual aspects of the procedure and the
choice of donor. It was in the early stages, before the referral,
that many couples needed help and advice. They were aware
that the decision to accept AID was more than a medical
decision, and many couples discussed with each other how
AID might affect their marital relationship, but in most cases
they were left to do this alone and were given no help in
identifying and exploring possible areas of difficulty and
conflict. Many couples would have appreciated guidance at
this stage; wife 1112 said, 'I think there should be somebody
to go to and talk with – as we're talking to you now – that
they might raise points that you perhaps haven't actually
thought of yourself.' Several couples remarked that the
interview had caused them to think about issues which had
not previously occurred to them.

Not all couples had the sort of relationship in which they
were used to talking out problems with each other and they
found it particularly difficult. One wife said, 'We don't
always talk about problems anyway . . . I needed someone
to talk to but you [husband] wouldn't.' Her husband explained, 'It's very difficult to talk to each other about it
without getting *too* emotional about it. You need somebody
that's outside.' Most couples felt that doctors would answer
questions of a medical nature but they were hesitant to make
too many demands on a doctor's time. Besides, it was not

answers to specific questions which these couples required but the opportunity to discuss and explore areas of doubt and uncertainty with a sympathetic and understanding counsellor. Some couples would have found it particularly helpful to talk to other couples who were having similar problems but this was impossible as most did not admit their AID involvement to others. Even couples who met in the waiting room tended to feel constrained and unable to talk to each other. Wife 913 said, 'We certainly never knew for sure that anybody else was there for the same reason that we were there. And we felt that, although we would have perhaps liked to have chatted, it might have been embarrassing . . . We felt it was better for everybody if we kept to very general terms.'

Nevertheless, the majority of couples tried as best they could to make a carefully-thought-out decision for themselves, though not all couples were so conscientious. Some couples gave only superficial thought to the matter; one wife said, 'It was just a case of – that's it, we've got to get on with the next best thing.' Other couples too felt there was very little choice in the matter. Wife 609 said, 'There was no alternative. We didn't want to be childless. Adoption was out of the question. This was the only thing left. We just jumped in with both feet and hoped that it would be alright. But obviously you think through these things; we are realistic people, we realised it wasn't a bed of roses – you have to think about what might happen and be prepared.'

Many infertile couples feel a sense of helplessness and lack of control over their own lives when they learn of their infertility. The ability to have children had always been taken for granted by the couple; the only decision to be made was when to have them. When infertility becomes evident the couple can no longer implement their long-term plans for family life and they may feel that they are no longer in control of their own destiny. In a few cases this feeling of lack of control over their own lives spilled over into their commencement of AID. One wife said, 'We were referred from one doctor to another . . . We were totally dependent on the path that we were pushed down. The appointment was made and we just sort of fell into it.'

All the couples stressed that the decision to employ AID must be a mutual one with the husband and wife in agreement; as wife 906 said, '. . . because obviously I think it's something that unless you're both for it one hundred per cent then it's not something you should do.' Each spouse tended to view the situation from the perspective of the other. Husband 877 said, 'We were both feeling each other out really, because I did not want to hurt her feelings and obviously she did not want to hurt mine.' Later he explained, 'My wife said was I at all upset about it being another man's, a donor. And also I had to ask her a straight question – did *she* mind it being another man's baby inside her?' It was comparatively rare for a husband to appreciate that a wife may have feelings of rejection, either of the husband or the child, because it was not her husband's child she was bearing. Most husbands, when attempting to see the situation from the point of view of the wife, thought more of their wife's 'maternal instinct' and natural desire for children which was being frustrated by the husband's infertility. Husband 975 said, 'I think a woman's maternal instinct is very strong and I thought it would be very selfish of me, because I couldn't father a child, to deprive [wife] of her part'. All wives were aware that their husbands might have reason to feel excluded and jealous; one commented, 'To me I was getting everything. The baby was going to come with me; it would be more me than [husband], and I felt it was an awful lot to ask somebody. It seemed so unfair.' This meant that some wives, who had already decided that they wanted AID, felt unable to voice their decision but thought it best to wait and hope that their husband would take the initiative. One wife speaking to her husband said, 'I didn't say anything because I wanted you to say that *you* wanted it. And then when you said you

85

didn't mind, I said 'Oh, yes, I'd be quite keen'. Some wives tried subtly to lead their husbands to the 'right' decision. Wife 501 suggested to her husband that they should try for adoption, but confessed 'I think I wanted *him* to suggest it [AID], but I wouldn't do the suggesting'. Many couples also came to a tacit understanding that in an argument any mention of AID must be absolutely taboo; it was far too sensitive and potentially destructive a subject to be used as a weapon with which to hurt an opponent in a quarrel.

COMPARISON OF AID WITH ADOPTION

When making the decision to have AID, most couples had also considered adoption as an alternative, and 23 couples had made formal application to an adoption society before they began AID. Nevertheless, almost all the couples expressed a belief that AID was preferable to adoption; only 3 couples failed to express this point of view. AID was seen to meet the needs of an infertile couple in ways that adoption did not. The primary purpose of adoption agencies in arranging an adoption is to meet the needs of a child for suitable parents. But most couples were not applying to become adoptive parents in order to provide for children who were in need of parental care and family security. Indeed, they appreciated that special qualities were needed by adoptive parents if they were to be successful. The couples' primary need was to have a baby of their own and they saw adoption almost entirely as a means of alleviating their childlessness and meeting their own needs, rather than as a means of meeting the needs of a child.

Some of the reasons given by couples for their preference for AID expressed the negative aspects of adoption. Waiting lists of prospective adoptive parents were long, and the couple might have to wait several years before a baby could be offered to them; indeed, it was unlikely that a newborn baby would ever be available and far more likely that the

couple would be offered an older child or perhaps a handicapped child. Some waiting lists were so long that they were
now closed to new applicants. Couples also resented the
selection process to which prospective adopters were subjected and felt that far too much emphasis was put on
material standards of living, housing conditions and so forth
rather than on how suitable the couple themselves were to be
loving and adequate parents. Couples were unhappy about
the initial period of uncertain custody of a child before the
adoption procedure was completed; they were afraid that
they would be unable to commit themselves fully to loving a
baby as their own if there was the possibility that the baby's
mother might change her mind and reclaim the baby. Some
couples were worried that an adopted child might inherit
undesirable characteristics from its unknown natural parents
and might become troublesome and difficult to control as it
grew up.

But couples were also aware of positive advantages in
opting for AID. Perhaps the most common comment was that
the child would be more 'theirs' as it would be part of at least
one of them. The wife would be able to experience pregnancy
and childbirth. Not all wives felt an urge to experience
pregnancy and bear a child themselves, but in some women
this urge was very strong; one wife said '. . . because I
wanted to be pregnant. I thought if I don't get pregnant I
won't be complete.' Another mother, who had an adopted
as well as an AID child, said '. . . to bear a baby, to be pregnant
and have a child. I have been a different person since I had
the baby – far more fulfilled. I needed to feel a woman.'

Because of this pre-natal and birth experience, couples perceived that the bond between parent and child would be closer
and stronger. Husbands could also share in this experience
and the majority of husbands were present at the delivery of
the child. Several wives were aware that this would increase
the husband's feeling of participation in the process and was
likely to strengthen the bond between father and child and

so encouraged their husbands to be present. Wife 901 explained, '[husband] is a bit of the squeamish type, and I said well, if you could just stay with me in labour, and then if you don't want to see the birth . . . I was in labour a long time and when it came to taking me to the delivery room they just stuck a gown on [husband]; so they didn't really give him an awful lot of choice! But he said he wouldn't have missed it for anything. In fact he cried a little bit.' The husband continued the story: 'I thought, I don't even like watching it on telly! It was only when it was over, and they suddenly held the baby up, and he opened his eyes for the first time. I can see his face now – it's a wonderful experience.'

Wife 973 explained that it was not that she had a great desire to experience pregnancy as such, but '. . . I looked after a friend's baby for a week once, and it was about a year old then and it kept doing things, and I kept thinking "I wonder if *my* baby would have done that at that age?" And I thought if I adopt a baby I'm going to spend the rest of my life thinking that – saying I wonder if *my* baby would have done that.' Another mother, who was herself an adopted child, and who had a very good relationship with her own adoptive parents, felt she wanted to bear a child herself because she felt a need for a biological link which would give her a sense of reproductive continuity. She said, 'I wanted a baby of my own, because not having a mother of my own I wanted a baby of my own.'

Some couples felt AID was preferable because the genetic origins of the child would be known to a greater extent than in adoption. Wife 906 explained, 'With adoption, you don't always know much about the parents. With AID you know it is half mine, and you know all the donors are carefully vetted.' Later in the interview she expanded this comment: '. . . some adopted children are from teenage mothers, perhaps with not very good backgrounds, and you don't know how that child might turn out. Because I know a lot of it depends on environment, but it is partly heredity, and I think in a way you can feel more sure with AID than with adoption that you are likely to end up with a reasonable outcome.'

Couples were aware that if they were to adopt a child they would be taking over responsibility for an existing child who had originated from outside their marriage; but if they were to have a child by AID it would be a child resulting directly from a decision made in their marriage – a *social* creation of their own marriage even if it were not a genetic procreation of their own. Couple 501 explained that, if it had come to it, they probably would have decided against adoption. The wife said, 'I think it was because we both desperately wanted a child that was *ours* right from the very start.' [Her husband interjected, 'Part of *our* unit'.] 'We didn't want somebody just to come and knock at the door and say "Right, we've got a baby for you."' Husband 1112 explained the difference very well: 'AID is an essentially different procedure from adoption. It's the act of creation, whereas adoption is a sort of accommodation.'

AID also allows a couple to behave as though their baby is the result of a natural conception, and so to pass as a 'normal' family. Husband 609 said, 'The very fact that you get pregnant the same as everybody else, the baby comes out the same way as everybody else, you're the same as other people and that *is* important.' If family and friends all assume that the new baby is the result of a natural conception, this means that the child also need never discover his different mode of conception. Couples perceive many differences between AID and adoption, hence they are reluctant to accept the suggestion that adoption practice and procedure (particularly with regard to legal registration, and informing the child) should serve as a model for AID practice.

AID ANXIETIES AND EXPERIENCE

Having made the decision to have a baby by AID, couples did have some anxieties about the implications and the outcome of their decision. These were not major, concrete worries but shadowy, vague fears about things which *might* happen. At the end of each interview, when couples were asked if they had any specific questions of their own, the most common first question was to inquire if other couples had problems. Most couples appear to suspect that AID is likely to cause problems even though their own experience may not have suggested this.

The major preoccupation of most couples was to ensure that people outside the medical profession should not discover that they were having AID. There was a common feeling that other people would not understand how their overwhelming desire for children drove them to meet this need by going outside the marriage relationship. Wife 863 said, 'They might not understand the need that we had; what we went through to get them. Some people have children quite easily, with no difficulty at all – they might not see the need to go elsewhere.' Many couples believed that their parents, because they were of an older and more conservative generation, would not understand and might also disapprove of their action.

Couples were slightly anxious about how they would feel towards the child. Wives tended to worry more about their husband's reaction than their own and did their best to involve the husbands in the preparation for the birth. Some husbands were confident that there would be no problem, but many confessed that they were anxious that they would not be able to accept the baby as their own and that they would not have a spontaneous surge of pleasure, pride and love at the birth. In all cases these fears proved unfounded and the husbands who had been present at the birth of their children had experienced profound and positive emotions. But some couples were aware that to love and accept a baby or young child is one thing – to maintain that acceptance throughout childhood, and particularly the period of adolescence, is another. Husband 902, a thoughtful and articulate man, whose love for his two AID children was self-evident, said, 'That still does concern me, actually. I suppose it's the point of concern that adoptive parents have, that when you get to adolescence and the balloon goes up as it were, it's easy to shelve off responsibility by perhaps even in anger actually saying "Well he isn't mine anyway. This is nothing to do with me." And it does worry me that one hopes one is going to have the equanimity and stability to weather any storms of that kind.' A very few mothers admitted that when their children were fretful or difficult they felt a need to protect their husband from this disruptive behaviour. No doubt many mothers of naturally conceived children feel this need too, but AID adds an extra dimension of anxiety. One soldier's wife said, 'I must confess that occasionally it's passed through my mind – when my husband has been away and comes back from a trip somewhere – and this monster, crying all the time – I feel I must get him [child] out of the house. Also I feel that [husband] must think, oh he's not really mine. I feel I should shelter my husband slightly, [she laughed] which is silly really. But I do feel that. I should protect him.'

Although these fathers had accepted their children the concept of paternity still caused confusion. It was noticeable that in trying to explain their feelings about fatherhood, very many of the husbands referred to the donor as the 'real father' although they were in no doubt that the child was 'their' child. Husband 878 tried to describe his confusion. 'The fact it was not mine – they are, I know they are mine – but it was in the back of my mind that it was never mine, never my child – I wasn't the one, the father – but I mean I know they are mine now, they always will be mine . . .'

Couples also had some anxieties about the qualities of the donor. The donor would remain anonymous and so the couples realised that they must trust the choice of donor made by the practitioner. They were concerned that the donor should be healthy, but presumed that as AID was provided by the medical profession all the donors would be medically examined and fit. Some were also concerned that the donor should be reasonably intelligent. Husband 920 summed it up by saying, 'I was pleased to find that the donors were vetted thoroughly and were physically sound and usually fairly intelligent people. I know the nature/ nurture debate goes on but its nice to know you are covered both ways.' But the main concern of most couples was that the donor should resemble the husband in physical characteristics; couples were very anxious that the child's appearance should be compatible with their own.

It was apparent that many wives were hesitant to admit to having any thoughts at all about the donor for fear of hurting their husbands' feelings. Wives who did admit to thinking about the donor described an attitude of fairly superficial curiosity rather than a desire to know who it was. Wife 975 wondered '. . . what he did for a living, what he looked like, was he clever, how tall was he–all these things. But I wouldn't really want to know. It's just my curiosity. It would bring it too much to reality. I would be thinking "I've had your baby."' Some couples felt they ought to know more about the donor, not his identity but his aptitudes and talents. Husband 1015 said, 'I think we should know a bit more about him [the donor]. I mean, I don't like football or rugby or anything like that, but I'd like to know whether the donor does, to know where to channel [son] to. He [son] likes reading books and he likes writing or drawing; was his father a poet or a writer? He likes band music; does his father play a musical instrument? I don't; I wouldn't channel him into music because I'm not interested, but if I knew his father was then we could do something about it. I don't want to know *who* the donor is, or what he is, or where he comes from, but we would just like to have known what he was good at.'

There was also some concern about how frequently the same donor was used. Several couples were anxious about the possibility of inadvertent intermarriage between half-siblings. Many couples, when trying for a second child, expressed a preference for having the same donor again. This was partly because they were happy with the characteristics of the child they already had, but also because they felt that the siblings would grow up to be closer to each other if they were full 'blood' brothers and sisters rather than half-siblings. They wanted to approximate as closely as possible to the 'normal' family situation and reduce additional genetic links to the minimum.

There was very little awareness by couples of the donor's viewpoint. Husband 913 said, 'I hadn't thought of it from the donor's point of view at all. There is a complete blank. The boundary of the problem stops at us – we look inward.' Wife 878, towards the end of the interview, said, 'God – it must be weird for a man who has donated – and not knowing how many children he could have floating around. I'd never thought about that.' A few couples expressed doubts about what would motivate a man to donate semen and a few couples expressed gratitude to those men who did, but the main impression was that the couples were completely un-aware of the donor as a person who had his own involve-ment in the AID process to consider.

Couples were not anxious about the legal situation. They realised that technically the children should be registered as 'father unknown' but, without exception, all the husbands had entered their own name as the father. Husband 913, a solicitor, said 'I couldn't see that it made any difference. I based that on my own understanding of English law, that it is based on the tenet of what is fair and reasonable. And I thought that what we were doing was fair and reasonable.' A more typical comment was that of couple 895: 'We thought, "If it's against the law – hard luck!"'

SECRECY

Almost all the couples felt a need to keep their AID involvement secret. Only 2 couples had told all their family and allowed friends and neighbours to become aware, and another 4 couples had told all their immediate family but no friends. Thirty-three of the 57 couples interviewed had told no-one at all outside the medical profession. The remaining couples who did not keep AID a complete secret, selected out certain relatives or friends whom they would tell. Ten couples told both sets of parents and 7 couples told one set of parents but not the other; 2 couples told the wife's mother only. Fourteen couples told some selected siblings but 4 of these had not told their parents. Eleven couples told selected close friends, and 2 of these had not told any relatives.

Without exception all the couples who had told relatives and friends had found their confidantes to be understanding and accepting, and the couples found it had been helpful to share this confidence. Husband 892 said 'Yes, I think they ought to know, really. I think it definitely makes it easier.' His wife continued, 'I couldn't bear the thought of people coming into the hospital – not so much my Mum and Dad, because it didn't affect me – but [husband's] Mum and Dad coming in and saying oh, it's got [husband's] nose or ears or whatever, and I couldn't go through that – the lies.' Her husband added 'And I think, at the time, as well, we just wanted somebody close to talk to about it. That came into it a lot. Because we had been through such a drawn-out process. I find it helps me to talk about things like that to other people.' With the exception of two paternal grandmothers, all the grandparents who were aware that their grandchildren were conceived by AID had accepted these grandchildren whole-heartedly, and with the usual pride and delight. Wife 1050 remarked, 'The grandparents treat the child just the same as any other grandchild – they thoroughly spoil him!' The two exceptions were reported to be rather uninterested in their grandsons, but in both cases there appeared to be other underlying family conflicts.

Only 3 couples had decided to tell the child of his or her

AID conception when the child was older. Forty-eight couples were certain they would not tell the child, and 5 were undecided. In one case the husband and wife were in disagreement about the issue, but the wife had promised to observe her husband's wish that the child should not be told.

COUPLES WITH GROWN-UP AID CHILDREN

Fifteen parents who had AID children who were all over the age of 18 years were successfully contacted, and 11 were willing to be interviewed. This sample of older couples shows bias in that they had all voluntarily kept in contact with the practitioner over many years. Nevertheless, some couples in this group had experienced quite serious problems, including divorce, therefore it was not the case that only 'successful' couples had continued to maintain contact. Nine interviews have been completed and one couple who lived too far away to be interviewed wrote about their experience. In four cases the mother only was interviewed; three mothers had been widowed and one was divorced.

At the time of AID all these couples told no-one outside the medical profession of their decision. The 10 couples produced a total of seventeen children, all now alive and well, the eldest being 40 years old. Some of these children were married with children of their own, and all were well loved and had brought great happiness to their parents. In most respects the story told by these couples was remarkably similar to the experience described by the younger couples. The older couples had in a very real sense been pioneers in accepting a new and controversial procedure which had met with considerable criticism both in parliament and in the popular press. They imagined that couples these days must have a much easier time; they were surprised to learn that it was still difficult to uncover information about AID and that there was still a perceived lack of acceptance of the procedure, and that young couples today were still experiencing many of the difficulties which they themselves had experienced twenty or thirty years previously.

The main difference in the findings of the interviews of older couples was that, contrary to expectation, several children had been told that they were conceived by AID. In seven families AID had continued to be kept a complete secret and the children had not been made aware of their different origins, but in three families the children when they were grown up had been told by their parents that they were conceived by AID. The experience of the older parents does seem to indicate that there can be a change in attitude as children grow older. Husband 849 said, 'While they were young I was only too proud and pleased to have felt that the children were mine, and nobody *ever* inquired or said "They're not your children" or anything like that. So all the time they were juniors I felt confident that nobody would query it. It was not until the later date that my son came up with this at the age of 14, that I couldn't be his father, that it all started ticking over. And I thought, well, eventually I will have to tell them.' This son had decided, following an 'O' level biology lesson on genetic inheritance, that his own eye colour and the eye colours of his parents did not fit with the rules of dominant and recessive genes which he had just been taught. Although this boy was not told of his AID origins at the time of this query, he was told a few years later and it is likely that this episode had some influence on his parents' decision to tell him.

A clear pattern emerged among the three older couples who had told their children. They had done so because they perceived that the children had a problem which could be alleviated by allowing the child access to the minimal knowledge which the parents had about the donor. In one case an AID son, whose father had become an invalid in later life, was about to marry. The mother perceived that the son, and his future wife, might worry that their children could inherit some physical disability. She knew that all donors were required to be healthy and could therefore reassure her son by telling him this. Another mother was aware that her son's wife doubted his fidelity because his father had had extramarital affairs. This mother told her son of his AID origins because she felt certain of the moral rectitude of the donor. She said, 'I knew this lady doctor wouldn't have picked just *anybody*'. She then felt the son need no longer worry about inheriting undesirable moral traits from her husband. A third couple were disturbed about the lazy and rebellious behaviour of their academically bright, younger AID son who had dropped out of university. His father was a manual worker, but the couple had been given to understand that donors were professional people. They therefore told their son, so that he should know that he had the potential of a high intellectual capacity in his inherited characteristics. His elder brother was then also told that he was an AID child.

THE EXPERIENCE OF AID CHILDREN

Hearsay evidence of the reactions of these four young adults to knowledge of their AID origin was heard from their parents. In another family the parents had divorced, the AID daughter going to live with her mother. When attempts were made to contact this mother, a reply was received from her daughter explaining that her mother had recently died. This young woman guessed the reason behind the contact as she had been told of her AID status by her mother, and she consented to be interviewed. Two other young adults contacted the research team; one was interviewed and one wrote of her experience. It is interesting to note that these three young people had also been told because it was perceived that they had a problem which could be alleviated by telling them of their AID origins. The experience of one young woman who had a father who suffered from depression and who was told just before she married, fits exactly the pattern already described. In the case of the other two young people, knowledge of the donor's characteristics was not an important factor. One was bothered by the behaviour of her (now divorced) father who wished her to visit him after many years of lack of interest or contact. The other had become psychologically disturbed during adolescence and had needed psychiatric help.

These seven young adults (who had been told in their late teens or 20s) accepted their AID status equably. One young man said, '. . . It even surprised me that it didn't upset me particularly . . .' Another young woman said, 'I was sort of pleased in a way – being one of the first people to be born in this way by AID.' They were enjoying life and happy to be alive and realised that they owed their existence to AID. They were pleased to feel that their parents had wanted a child so badly and that they were that child which had fulfilled their parents' wishes. One said, '. . . the realisation that I'd been brought into the world – you know – they actually went to *tremendous* lengths because they wanted to have a baby. And I suddenly felt that they must love me a tremendous amount, that I was very important to them.' In the cases where the father was still alive and in contact with the child, the fear that the father/child relationship would be damaged proved to be unfounded. Indeed, the relationship appeared to be enhanced as the son/daughter came to realise the anguish which the father must at times have experienced. One young man said, 'To me he's my father. And realising the trauma they must have been through, having this huge, great secret, trying to keep it from me – you know I have tremendous respect for them.'

Some, but not all, the children had suspected that something was amiss; two had queried their eye colour, another had discussed his doubts about his paternity with an adopted sister, another had queried his blood group. But this young man also said, 'It was just a general thing – it was as if I'd always known there was something wrong, I'd always known there was something amiss – and suddenly, being told that, it was as if a huge great weight had been lifted off my shoulders.' Other research has shown that adopted children, who suspected that they were adopted, felt unable to question their adoptive parents, and this finding appears to hold true for AID children also. Even after these young adults had been told of their AID status, their parents reported that they did not again discuss the subject together. One young woman who had been told by her mother said, 'The extraordinary thing is that although we were so close I could never bring myself to quiz her more.' It may well be that older couples who believed that they had successfully kept the matter secret over many years were mistaken; their grown-up children may well have had suspicions which they felt unable to voice. One older mother said, 'Well, I mean, I'm not saying that the children might not have inklings; I mean, who's to say? I don't think perhaps they'd broach the subject with us. But they must put two and two together; eleven years [of marriage before the first baby] is a long time. And my husband has said to my son that we never used any contraceptives – so perhaps they have put two and two together. Who's to say?'

From the experience of these older couples and of the young people who are aware of their AID conception, it would appear that it is not only possible to tell an AID child of its origins, but that it is possible to do so without harming either the child or family relationships. However, the group of older parents was relatively small and it would be unwise to attempt to generalise from the experience of these small numbers of people.

In the next two chapters a closer examination is made of the ways in which infertility and AID affect the lives of these couples; in particular the concepts of secrecy and stigma, and the ways in which they contribute to the stress surrounding infertility and AID are examined.

99

6

AID and Secrecy

The practice of AID has always been shrouded in secrecy. AID practitioners have only rarely given explicit advice that AID should be kept secret, rather there has been an implicit and unquestioned assumption that secrecy is necessary and beneficial. In the early reports of AID the assumption of secrecy is often coupled with the discussion of the legal aspects of AID, and it is likely that the lack of a legal framework in which to practise AID was a major factor underlying the perceived need for secrecy. The legal registration of children born following AID poses a problem; strictly speaking they should be registered as 'father unknown' and are technically illegitimate, but in practice they are registered as the children of the infertile husband. AID practitioners advise couples to continue to have sexual intercourse during the period of AID, as where a husband has had access to his wife within the twelve months prior to the birth of a child, that child is regarded as legitimate until proved otherwise. In this climate of legal uncertainty it is understandable that secrecy was thought, both by doctor and patient, to be prudent. Other advantages of secrecy were also observed; the AID child need never learn the circumstances of his unusual conception and so could be protected from knowledge of facts which the child might find disturbing. Secrecy would also protect the feelings of the husband; his infertility would not be exposed as everyone would presume that the child was a 'natural' conception.

But it has become apparent that this stance of secrecy also has disadvantages. Secrecy restricts the availability of information to couples who could themselves benefit from AID if only they knew enough about it. Some couples who were interviewed confessed that they had friends who were childless and whom they suspected might benefit from AID, but they were reluctant to broach the subject for fear of exposing their own involvement in AID. Secrecy also implies that there may be something improper or even shameful about male infertility and AID which must be hidden and cannot be talked about openly. This means that some family doctors are reluctant to present this option to their infertile patients, or the doctors themselves may be disapproving of the procedure. This rather furtive atmosphere may also be one factor in restricting the availability of AID provision within the general health service so that couples, if they do hear about it, often have to travel long distances to an AID clinic. Resources for infertility services are not generally given a high priority but it is likely that the secretive nature of AID has also contributed to its restricted availability.

Secrecy hinders a public discussion of the procedure and so people remain ignorant of the many complex factors involved; this ignorance contributes to a lack of understanding, and attitudes within society remain prejudiced. It also means that members of the helping professions such as social workers or marriage guidance counsellors are themselves ignorant of the problems to a large extent, and so unable to offer expert help and counselling. Secrecy means that pressure is not put upon the legal system to institute necessary legal changes, nor is an agreed regulation of the practice possible and so the procedure remains open to abuse.

Perhaps the most important disadvantage of secrecy is that it deceives close family members and even the child

101

itself. As we have already seen in Chapter 3, the implications of family relationships are far-reaching. Relatives are aware of ties with each other from which arise duties and obligations and also rights. Secrecy which deceives family members about their ties with close relatives undermines the basis of trust on which all family relationships are founded. Secrecy also deceives the child about its own identity and genetic history. This deception is no doubt carried out with the best of intentions and in order to spare the child distress; nevertheless, paternalistic lying has its own dangers and closer examination often reveals elements of self interest on the part of the person who is withholding information. Besides, the perspective of the deception from the point of view of the person being deceived is often quite different; he usually feels wronged and let down and suspicious. In short, he has been taken for a ride. In the early days of AID it was suggested by one practitioner that it might be kinder not to inform the husband of his sterility and to perform inseminations without his knowledge and so allow him the pleasure of assuming he had made his wife pregnant. This suggestion received the critical reaction it deserved, yet the deception of the child in a similar paternalistic manner has not provoked a similar hostile reaction. Not only are the children deceived about their identity, they are also misled about their genetic history. This means that if they are asked to provide a family medical history in case of illness they are either unable to do so, or unknowingly give a false history.

If the secrecy which surrounds AID is examined more closely it can be seen to have at least three component parts: i) the confidentiality of the consultation between the doctor and the infertile couple; ii) the anonymity of the donor; iii) the pretence by the couple that the AID child is the result of a natural conception. The confidentiality of consultations with members of the medical profession is essential and entirely proper. There would also seem to be good reason for maintaining the anonymity of the donor. If the identity of the donor were known, not only would there be legal complications but conflicting emotional ties between the family of the recipient and the family of the donor would also be likely to arise. Nevertheless, it has been suggested that certain non-identifying information about the donor could be shared with the recipient family which need involve no breach of anonymity, for example, a description of his interests, aptitudes, occupation, appearance and tempera- ment. It has also been suggested that a register of donors should be kept so that non-identifying information relating to the donor's medical history could be divulged to the recipient family in case of need. But it is the third component of secrecy, the purposeful misleading of close family mem- bers and of the child itself which gives rise to most social concern.

From the evidence of the sixty-six interviews conducted, it would seem that keeping AID secret is a practical possibility. In adoption, relatives, friends and neighbours are all aware that a child has been adopted by the couple, but in AID the wife becomes pregnant, apparently in a normal way. Among the older families, some couples had kept AID secret for as long as thirty to forty years and were convinced that it still remained a secret.

Many couples were reluctant to use the word 'secrecy' to describe their behaviour and there was considerable dis- agreement about a definition of the concept of secrecy. Several couples denied that they were keeping a secret at all and saw their behaviour in terms of *privacy*. Husband 902 said, '. . . I think it's so private and personal, like conception itself, that really it's a personal confidence you don't share with anyone else'. Whilst secrecy and privacy may both be concepts on the same continuum describing behaviour or knowledge which is not shared, privacy (in contrast to secrecy) is usually seen as a good thing, a state of affairs which must be defended and preserved. Husband 461 said, 'I think we have chosen not to tell people because it's none of their business; why should you tell them personal details?' 'Privacy' describes an area of personal autonomy separated off from shared experience, and clearly there are large areas in each person's life which are theirs to keep as secret as they wish. However, certain behaviour or knowledge may have implications for others, and others may have a stake in learning about facts which will have an effect on their own lives. Whilst an infertile couple may consider the fact that their child was conceived by AID to be a private matter, the grandparents of that child (if they become aware) may well feel that they had been unfairly excluded from knowledge which directly affected them.

Some couples saw their decision not to share knowledge of AID with other relatives as a matter of *propriety*; sexual relationships were not to be talked about and AID should be treated in a similar manner. One husband said '. . . It's a personal thing. I wouldn't go to my father and say "I made love with my wife last night." I feel there is a similarity.' Other couples saw their decision not to tell close relatives about AID as a matter of *diplomacy*; knowledge of AID might cause strife in the family, and to keep such knowledge to themselves would prevent family upsets and unhappiness. Most couples do not wish to admit, even to themselves, that they are being secretive or deceptive; they attempt to justify their behaviour in more socially acceptable terms.

MEANS OF MAINTAINING SECRECY

There is on the part of most couples a desire to avoid outright lying; couples try to restrict themselves to white lies. This point is clearly illustrated by one wife who said, 'I had to explain to my parents that I had seen a specialist for *my* infertility problem. This has proved very handy in parrying awkward questions from well-meaning relatives and others. My periods are haywire anyway, so I feel I've got some justification.' It was a common ploy for the wife to pretend that she was the infertile partner. This provided a very plausible explanation as it was the wife who was paying monthly visits to the gynaecologist. The wife had also to keep a daily early morning temperature record to indicate the time of ovulation which must be predicted accurately if AID is to be successful. If conception did not readily occur the wife may have been subjected to investigations to examine her own fertility, and she may indeed have needed treatment. It is easy to see why for many couples the sense of deception becomes blunted as the wife gradually comes to see herself as the patient.

Some couples maintained secrecy, and at the same time salved their own consciences, by a partial telling and a partial withholding of the truth. For example, some couples explained to parents that they were having artificial insemination but that it was the husband's semen that was being used. Other couples avoided actively telling a lie, but in a passive way just avoided telling the truth. Couple 755 said 'We never really *decided* it, we just never bothered telling anyone . . . No-one seemed to ask, or anything like that.'

A major factor determining whether couples are secretive or open about AID is the need to maintain a stance which is consistent with previous behaviour. It seems that where a couple have initially been open with their parents they tend to continue to be open and discuss the results of fertility investigations and possible solutions to their childlessness. This may simply be because if a couple have been open with their near relatives, they must maintain a consistent story and the option of secrecy is less open to them. Similarly, where a couple have decided to conceal the husband's infertility from close relatives, secrecy about AID is necessary in order to continue to keep that knowledge hidden. Once a couple have started to keep AID secret it becomes progressively more difficult for them to tell relatives even if they later wish to. They would then have to admit to having been

105

secretive and to having withheld information for some time as well as to admitting involvement with AID. One wife commented about her own and her husband's parents, 'I think if we had told them straight away perhaps they wouldn't have minded; but now I think they would think "Why couldn't they have told us earlier?"'

REASONS FOR SECRECY

One of the commonest reasons given for secrecy was merely a denial that there was any reason to tell. Husband 973 said, 'I keep on coming back to the fact that because it's successful I can't see any reason at all that anybody else needs to know.' This belief that there was no reason to tell was closely linked to the desire of the couple to present themselves as a 'normal' family. As one wife put it, 'We just want to carry on as though nothing has happened'. Secrecy allows a couple to deny AID. Another husband explained, 'I've got photographs of the baby being born, and I've got cine-films taken from when they came out of hospital all the way through. To all intents and purposes the children are mine and that's the way I'd like to keep it.'

This desire to appear as a 'normal' family points to the main reason why most couples think it best to keep AID secret. The most common reason given for secrecy was fear of stigmatisation. Husband 1030 said, 'I would find it very difficult to tell people, because it isn't so much the way I feel, I think it's the way other people would react. You can't rely on other people to be as thoughtful and understanding about problems as – well – as I am about my own problems.' If their recourse to AID were known, many couples fear that their standing in the eyes of other people would be diminished and that they would in some way be discredited. Couples expressed particular concern about the child, that it also might be labelled as 'different' and stigmatised. But experience of adoption suggests that adopted children are accepted by most kin and are not stigmatised, and it seems likely that AID children would fare similarly. The over-whelming impression gained at interviews was that the paramount reason for secrecy was the protection of the feelings of the husband. The concept of stigma plays a very important part in generating the behaviour which surrounds AID. This is discussed at greater length in Chapter 7.

Couples were aware that AID held important implications for family relationships, particularly in respect of grandparents. Their parents would expect that any grandchildren would result from within the union between their respective son and daughter. Couples were also aware that AID would create a situation of imbalance, with consanguineal or 'blood' relationships being present on the maternal side of the family which would be absent on the paternal side. Therefore some of the reasons given for secrecy involved the preservation of harmonious family relationships. Couples wished to protect the feelings of grandparents. Husband 501 said, 'It was largely to save them [grandparents] embarrassment. It was more thinking on what *they* would feel than on what we felt, because we are quite happy.'

When a couple marry, their parents generally look forward to the arrival of grandchildren and couples wanted to fulfil their parents' wishes by presenting them with their own grandchild. One wife explained that when they commenced AID she had said to her husband, 'I shouldn't tell anyone, and if I do get pregnant then all they know is, it's yours. And they were delighted because we'd been trying for six years. We went out there one evening and said I was pregnant, and they were over the moon. And his [husband's] grandmother, she's 82 and it's her great-grandchild, and she thinks the world of it because it's [husband's] you know. And I wouldn't say anything. If I thought they would find out it would break my heart.' Here again we have an example of a paternalistic type of deception; this time it is old people rather than children who are, unknown to themselves, being protected from the truth. However, it should be remembered that in the cases where the older generation were told, almost without exception they accepted the situation unreservedly, and rallied round with support, and derived tremendous pleasure and satisfaction from their AID grandchildren.

The decision not to tell grandparents was not only to save the grandparents distress; couples were also worried that their parents might disapprove of their decision to have a child by AID. Wife 313 said, 'I just don't really know which way they'd take it. My family aren't strict, they're far from that, but I think they . . . I suppose they've had children, Dad and Mum together, and they might have thought it was the wrong thing to do. If we couldn't have children we should go without rather than have somebody else, to produce it.' Couples were also fearful that if their parents knew, they might reject the grandchild and not consider him or her as properly part of the family; their parents might discriminate against an AID grandchild in relation to other grandchildren, and not value or treat him or her the same as a full blood-related grandchild. This was particularly so in the case of the husband's parents who would have no consanguineal link with the child. Nor did couples like the idea of maternal grandparents feeling that they had a stronger bond with the child than the paternal grandparents. Couple 609 were worried that the husband's parents might not feel able to accept the children as their grandchildren. The wife said, 'The babies would be no part of them whatsoever. It's very, very hard. How do you tell them that? They are bound to view them differently, especially since [husband's] sister had three boys which *were* part of them.' And so this couple, because they had not told the paternal grandparents, did not tell the maternal grandparents either. Again, it should be remembered that in almost every case where grandparents were told, they accepted their grandchildren and in no way discriminated against them.

AID and Secrecy

There was also a very understandable fear that if the child became aware that it was conceived by AID, the child's relationship with the husband would be damaged, and the child might reject him and no longer consider him as its father. Husbands were also fearful of what such a revelation would do to the security and stability of the child itself.

THE EXTENT OF SECRECY

Professional advisers and other authority figures were more likely to be told of AID involvement than friends or relatives. Couples perceived a more formal, distant relationship, and trusted professional standards of confidentiality. It seems that couples themselves saw a difference between confidentiality and secrecy. Doctors, midwives, health visitors, adoption social workers and even employers were told, whilst in all other respects the couple maintained a stance of total secrecy, and considered that the secret had not been broken. This trust, however, was not extended to the clerks and receptionists who worked in health centres or hospitals, and there was some worry that mention of AID in case notes might be seen and discussed by them.

In the sample of younger couples interviewed, 58 per cent kept AID a complete secret. Couples where the husband was a professional or white collar worker were more likely to keep AID a complete secret than couples where the husband was a manual worker; twenty of the thirty non-manual couples told no-one (66 per cent) compared with eleven of the twenty-five manual working couples (44 per cent). The number of couples who told both sets of parents was small, but manual working couples were twice as likely to do so as non-manual couples. It would seem that the relationship of these married couples is a very self-contained one, and the couples suffer or enjoy a considerable degree of autonomy and independence from their wider family. This self-containment is usually associated with the life-style of middle-class families and we have already noted that most AID couples were middle class in occupation or life style.

Apart from social class no obvious factor emerged to differentiate couples who told relatives from those who did not. Couples who had told their parents were likely to perceive a close family relationship and to live near the parents. But among couples who had told no-one, the majority also perceived a close family relationship and lived geographically near to their parents. Among couples who had told only one set of parents, the parents who were confided in were described as 'close' and lived nearby, but in three of these seven cases the parents not told were 'close' and lived nearby too. Because AID produces an imbalance in consanguineal relationships favouring the maternal grand-parents, it could be supposed that couples would be more likely to tell the wife's than the husband's parents, but this was not the case. In some instances the wife's parents were not told, even though the husband's parents were aware; this seemed to be because the husband was already disadvan-taged in some other respect. The wife's parents may already be unhappy at what they consider to be an unfortunate match for their daughter, and to tell about AID might exacerbate an already difficult situation. One wife said, 'A woman's family can feel that the husband has let her down, even if the wife doesn't feel that herself'. This illustrates an important point; infertility and AID must not be seen as isolated or discrete problems. The problem of AID is but one item in a complex of interrelating issues and family pressures which should be considered as a whole.

REASONS FOR OPENNESS

Besides concentrating on secrecy, we must also remember that more than 40 per cent of couples had told some relatives

or friends about their recourse to AID. Some had done so
because they felt relatives ought to know and it was natural
to tell them. One husband said, '. . . because we are very
close to them, and we've never had any secrets from them,
we didn't think this should be a secret from them. They are
part of the close family circle . . . So it just didn't dawn on us
not to tell [the] family; there was no reason to keep them out
of it.' Other couples told their family because they did not
want to become involved in subterfuge; one couple told
their family about six months after starting AID. 'It was
getting so involved. I felt awkward having to sort of
"wriggle", if you know what I mean. When I came back
they would say, "How did you get on? Did she come up
with anything?" and all this, and it got terribly involved.'

All the couples found that they received psychological
support from their confidantes. Some had told relatives for
this reason, for example wife 936, who said 'I'm very close to
my mother. We talk about it and it makes it a lot easier.' Her
husband added, 'You don't want to take all the burden on
your own shoulders'. Other couples felt unable to tell their
family and looked to close friends for this psychological
support. Husband 902 said, 'So we took [friends] into our
confidence, and they were very supportive'. His wife said,
'And we could talk to them about it'. Her husband continued, 'I think to have made the decision to keep it totally to
ourselves would have made it very difficult, particularly in
the anxious times'. The wife added, 'And they get as involved
in it as you do. They were *so* excited when we conceived,
weren't they? They took us out and bought a bottle of
Champagne!'

There was also a feeling among some couples that it was
not realistic in these days to imagine that other people would
not guess that the couple were having AID even if they were
not explicitly told. One couple who had told all their family
suspected that some close friends had also guessed that their
child was conceived by AID; the wife said, '. . . because they

111

are friends that we've had for years, they have known of all the problems that we had, and I often wonder – I mean, people aren't thick. They must read the same magazines that I read, and whether they think to themselves "Mmm – I'm sure it *is* AID" – but nobody's ever said. It's a thing which I'm sure has crossed our friends' minds. It's crossed mine, because I've got a friend who is in exactly the same position as I am. She was six years trying for a baby, then all of a sudden out of the blue she is pregnant.'

All the couples who had told family or friends had found that it had been advantageous to do so and did not regret telling. Sometimes a parent's initial reaction was one of shock but in almost every case the parents had quickly accepted the decision without any reservation. Couple 1213 described their parents' reaction to AID when the wife had told her husband's mother. 'She didn't say a word for fifteen minutes. We were watching television and we could see she was thinking, because we're a very close family. And then she got up and made a cup of tea. Then she talked to us about it in detail. She said she had to think about it for a moment but she was proud of us for giving it a try and she was behind us all the way.' The husband said to his wife, 'Your mother was shocked as well, wasn't she?' The wife explained 'I said, "We're thinking about AID, you do know what that means?" So she said, "I think so. You won't have that done, will you?" And then she thought about it for a second and said "Well, of course, really it's better than adoption" and then she rallied round. It was just the initial shock I think that we were doing something that they hadn't thought about.' They described how both sets of grandparents adore the child and how it had made no difference to the way they treat the baby. The wife said, '[Husband's] mother – if you're going to be personal, she knows it's not his child – she idolises the boy, doesn't she? We'll go out there for lunch tomorrow, and I'll *lose* the baby. It's lovely really.' Later, when discussing how secrecy is possible this husband reaffirmed his belief that to be open about AID is the best way; he said 'It [secrecy] must be harder actually. The way we are now, we're free at mind aren't we? We're free at mind.'

There was no congruence between the decisions to tell relatives and to tell the child. The majority of couples who had told relatives said they did not intend to tell the child, and some couples who intended to tell the child had not told close relatives. In many cases secrecy was not a carefully-thought-out and predetermined strategy – it was more a process, something that developed over time. Some couples would have preferred to be open about it all but were hesitant because they were unsure of what the reaction of their relatives would be, and there was no way of testing this in advance; once the secret was out it would be out for good. This meant that some couples had an ambivalent attitude to secrecy. One young wife explained, 'Well, my mum is pretty broad-minded but I don't know how far. I've slipped it in a few times that they do do AID down here, but I don't know whether she has cottoned on to it. It's her that keeps on saying that [AID son] walks like his dad, and I don't know whether she keeps saying it so she's trying to tell me she really knows, and she does if I know my mum, there isn't much she misses!' Later in the interview she added, 'If she asked me outright then I'd tell her. I'll probably tell her before she dies so that she knows; but not yet.'

COSTS AND BENEFITS OF SECRECY

A stance of secrecy undoubtedly confers some advantages. First, an AID child is held in law to be illegitimate; by hiding AID, this status is avoided. Secondly, couples perceived that secrecy allowed them to appear to others as a 'normal' family. But secrecy also caused problems. One husband said, 'It just shows, in a way, you start a deceit or a pretence right at the start and you have got to maintain it. It's like any white lie or fib, it snowballs. That probably sounds more sinister than it is in some way. But it is true, you know in your heart that you haven't told everyone the truth, and if there is one thing to my way of thinking that is a drawback, it is that with AID.'

Couples found it difficult to conceal monthly visits to the doctor during the process of AID. If the wife was working she often felt obliged to invent a plausible alternative reason for her visit which would satisfy the inquiries of her employer or the interest of her colleagues. Often husbands accompanied their wives on the AID visit and in some cases they too had to provide a reason for wanting time off work. One wife said, 'My husband had to take time off work. That's the most difficult thing about it; you do tend to get caught up in a lot of lies about where you are going.' Couples were very anxious that they should not meet anyone whom they knew on these visits. It is perhaps worth noting that from the sample of younger couples (i.e. all couples who had a baby between January 1977 and December 1980) not one couple came from the city in which the AID clinic was situated. In most of the nearby centres of population an average of three or four couples were interviewed; it seems possible that couples were travelling to a centre where they were unknown. Indeed one young wife interviewed had decided not to go to the clinic available in her home town. She explained, 'Well, I preferred going out of [home-town] because I was a bit worried about seeing people that perhaps I knew in town, or even knew them personally. It would have been awful to know that I had gone to the same place.' The difficulty of concealing monthly AID visits was exacerbated if couples were attending for a second baby, as they also had to make arrangements for someone to look after the first child. One wife said to her husband, 'It used to be difficult. Your mother used to have [child] sometimes and I would say I was going on a shopping trip to London or something. And I used to feel awful because I thought, what if something happens to me, and I would have to explain what I was doing down there.'

Many couples had been longing for a baby for several years and this became an all-consuming desire. If AID were not quickly successful, couples became despondent and depressed by their failure to conceive. Secrecy meant that they had no-one with whom they could share their problem, no-one whose shoulder they could cry on, and they were cut off from the support and comfort which family and friends usually give in times of trouble. Even in the joy of at last having the longed-for baby, secrecy still caused some problems. Relatives, friends and neighbours all feel obliged to comment on who the new baby is like, and this causes some couples to feel uneasy and embarrassed. One wife confessed 'I find I swallow, and have to give myself a talking to, that they are *not* suggesting anything, that they are not prying or probing – it's just a natural comment.' A few couples felt guilty at deceiving the grandparents about the conception of their grandchild; one husband said, 'We felt a little bit uneasy breaking the news [of the pregnancy] – we had been trying for a baby for all this time. We went down to tell my mother and [wife's] mother, and we were watching their faces, and it passes through your mind, I wonder if they believe us? Do they know we are doing AID? I wonder if anyone knows?'

Keeping AID secret means that a couple may suddenly and unexpectedly be confronted with a situation where they have to think quickly in order to cover any embarrassment or confusion. These may be small and insignificant matters of themselves, but they may loom large in the concerns of the couple affected. One couple who each had their own separate GP had been referred for AID through the husband's doctor and the wife's doctor remained unaware. The wife described a visit to her own doctor: 'Just after the birth of this second baby – he hasn't bothered before you see – he did the postnatal check-up and he said "What sort of contracep-

115

tive are you going to use?" And he could see I was looking a bit stunned, because I hadn't thought about it you see, that he was going to ask me this question. And I thought, "Really, this is ridiculous. Am I getting just *too* stupid?" But still, at the same time, I couldn't tell him.' Other wives mentioned that they were sometimes put on the spot by women friends wanting to discuss and compare their experience and decisions about contraception. Other wives described how they were taken aback when asked about family medical history when they took their baby for innoculation. One wife said she 'went all hot for a minute' but then realised the AID practitioner would not use a donor who was not healthy. Another young wife wrote to say 'Before one has a child people don't seem to ask when you're going to start a family, however, when you've had one we seem constantly to be asked when we are going to have another'. Then she added, 'I don't find it distressing but it's certainly tactless, and very difficult to answer.'

The anonymity of the donor (as opposed to secrecy about AID) also caused some problems. Because they did not know the identity of the donor some couples had a nagging worry that the child might inadvertently marry a half-sibling. Also, because they were in ignorance of the donor's medical history, couples whose children became ill worried that this history might have implications for the diagnosis and treatment of their child's illness. In these cases concern about their child's condition invariably overcame the desire to keep AID secret and the couples decided to tell the paediatrician about AID.

SECRECY AND THE AID CHILD

Almost all the parents of the young AID children had decided that they would not tell their children that they were conceived by AID. There was, however, among some couples a realisation that circumstances might change as the child grew older. Couple 782 had kept AID completely secret and did not intend to tell the child, but the husband said, 'I think if for some reason he ever did think there was something up, if it was on his mind, I'd talk to him about it'. Some couples were uncertain what to do. Wife 755 said 'I'm not too sure what I will do. I don't know. I'd probably discuss it with [husband] first anyway. That is a big problem. I wouldn't want to tell him actually – that's the truth of it.' After a long pause, she added, 'The problem will be when he is older, he may find out, we might have to tell him. We are pretty open with him, but you see it wouldn't mean much to him – it's not like adoption. Well, if he's brought up the right way, and we hope he is, it won't make a bit of difference to him.'

REASONS AGAINST TELLING THE CHILD

The most common reason given by parents was that there was no need to tell. Again this was consistent with a position of defensive denial of AID. Husband 780 said, 'We haven't made any plans to tell them. As far as we are both concerned they are ours . . . we probably wouldn't have thought about it [AID] again if you [the interviewers] hadn't been coming down.' The wife had been seen to be pregnant and had given birth normally; the child had a birth certificate which named the husband as the father, the child was considered by everyone to be the child of the couple, indeed the couple themselves looked upon the child as completely theirs. In short, secrecy was possible, and the child unlikely to find out accidentally. It was only if the danger of accidental discovery became acute that most parents would consider telling the child; otherwise, in their opinion, a hornets' nest of trouble could be stirred up, and quite unnecessarily.

Other couples gave more specific reasons. Some of these were to do with protecting the family members and relationships involved. Children might be distressed or hurt by the knowledge that their father was not their father in all respects. One husband said, 'No, I really think that would be cruel to the child, to bring her up to twelve or thirteen years as her dad and then say "Well, I'm not really your dad".' There was also a fear that children might feel uncertain of their own identity because of their anonymous genitor. One husband, whose own parents had separated shortly after he was born, said 'I feel that it could be traumatic to a child to know that it was conceived in that way. I went for many years myself wondering what my father was like, but at least I knew – people talked about him – my mother used very occasionally to say something – my grandparents talked about him – and I saw him briefly on one or two occasions. To imagine it was somebody who nobody *knew* I think could have a disturbing effect on a child.' Over and above the fear that a child's relationships might be disturbed, was the fear that a child might be generally stigmatised. Husband 1004 said, 'I think it would be very harmful in junior school – if anybody found out, children can be very spiteful. If she [AID child] knew about it at that age she wouldn't be able to keep it to herself, so I think it's best forgotten.' Couples were fearful that telling the child would also be hurtful to the husband. Wife 313 was adamant that they would not tell the children. 'Never, never. I think it would be awful . . . it would be the same as saying to them he's not really your daddy. No way – I just couldn't.' Couples feared that the child might think less of the father or even reject him completely, and the father/child relationship be spoilt. One husband said quietly, 'I'd be afraid perhaps he might turn against me – I'd be afraid of that.'

Other reasons were more pragmatic; a child might in turn tell relatives who had been excluded from the knowledge of AID, or even make the information generally available to schoolfriends. Couples claimed that it would be difficult, if not impossible, to explain AID to a young child – and if telling were left until the child were older and able to understand the technicalities, it might then be too late from a psychological point of view. Besides, there was no positive information about the donor which parents could give to the child, or which the child could find out for himself or herself from records, as in the case of adoption. Husband 1030 had obviously given the whole matter considerable thought: 'I don't see any *need* to tell her she's an AID baby. With adoption I think perhaps there is a need because our families would know and there is no way you can keep it from a child. I don't see the need to tell her. It would just confuse her. There's no way you could ever find the father. She wouldn't even know for sure that I wasn't the father. It would just, I would think, make her unhappy. There's no way you can broach it to a child to understand – and if you leave it until a child is old enough to know what it is all about, then I would have thought it would have been a real blow at that stage. It's very difficult. And of course you are being a bit selfish about it as well – you want her to think that I'm her father. There is no answer to some aspects of AID.'

Although couples emphasised that their main concern was for the child, a closer look at what was being said gave the overriding impression that the children were not being told because of the benefits this would bring to their parents, particularly the father. Many couples, after they had detailed what potential harm knowledge of AID origins could do to a child, also admitted that they thought their child would be able to cope. Husband 971 said, 'In our circumstances I don't think it would [cause harm] somehow. It would be a shock, but he gets so much love and attention from both of us that I don't think really it would make a lot of difference.' The withholding of information from the child was seen almost entirely from the parents' viewpoint. When the child, or indeed other relatives, were considered it was in a paternalistic way, and deception was justified by the claim that its purpose was to protect others from being hurt. Only rarely was it acknowledged that the other family members have

119

their own stake in this information which is being withheld, or that they have a right to information which is of direct relevance to them.

It would seem that rights, duties and obligations with regard to other family members are given a hierarchy of relative importance. Initially, a young couple may feel that the husband's right to have his infertility kept secret holds priority and the prime duty of the wife is to maintain the secret of her husband's infertility. The right of a small baby to such a complex and abstract matter as an accurate knowledge of his own origins can, parents feel, be safely and honourably ignored. But there is evidence from the interviews with older couples, that as the child grows older his or her rights begin to exert a greater claim for consideration and begin to compete with the right of the husband to secrecy. The growing young person's right to knowledge about his or her own origins may now be acknowledged. At the same time, the obligation on the part of the wife to maintain secrecy about the infertility of her (now ageing) husband may be seen to be reduced. This hierarchical view of rights and obligations within family relationships may help to explain also why so many couples keep AID secret from their own parents. Their rights as a couple and their obligations to each other are perceived to supersede the rights of, and obligations to, their parents; where there is a conflict of duty and obligation the rights of the parents are relegated to a position of lesser importance.

Only eleven couples believed that there was any chance of their child finding out about AID accidentally. The danger of accidental disclosure was seen to lie in two main areas; some couples were aware that reference to AID was contained in their child's medical records, and the child might one day inadvertently see this. Alternatively, a child may acquire sufficient biological or medical knowledge to query its paternity where there were inconsistencies in eye colour or blood group, or where the husband suffered from a pathological condition which could be associated with sterility. Couples who were aware of these possibilities still did not intend to tell the child unless accidental exposure became certain and imminent. The majority of parents who had told a few selected relatives or individual close friends about their recourse to AID, did not consider that this made it more likely that their children would find out accidentally. They were confident that the secret was secure and that knowledgeable persons would maintain the confidence.

REASONS IN FAVOUR OF TELLING THE CHILD

The most common scenario outlined by young couples in which they would tell the child, involved a fantasy in which the child was sterile himself, or in the case of a girl married to a sterile husband. A surprising number of couples explained that the only situation in which they could envisage themselves telling the child, was if the child found itself in a similar situation of male infertility. It seems that the parents saw AID as a solution to their own problem, and so if their offspring were presented with a similar problem the parents would explain how they themselves had found a solution. Because the parents saw AID as a *solution* it did not seem to occur to them that from the child's viewpoint AID origins could be an added problem. An alternative, and perhaps more likely explanation for the prevalence of this scenario, may be that if children were to share the couple's experience of infertility and need of AID, this would then bring them all together by reinforcing a feeling of solidarity and thereby increasing the likelihood of mutual support. A few couples were worried about the danger of half-sibling intermarriage, and thought it might become necessary to warn the child. The small number of couples who had told the whole family and some friends, thought it would be preferable for them to tell the child than for it to find out accidentally. One explained, 'I think we will have to tell her because of the fact

121

that our family know – because otherwise the chances are that it will come out, not intentionally, but it could come out. She should hear from us as opposed to hearing from someone else.' Only three couples expressed the view that the child was entitled to know, and it would not be fair to keep a child in ignorance of his or her AID origins. One wife explained, 'The Health Visitor said there is no real need to tell him, but I think I will have to wait until he is older to decide. I'd like him to know really. I think we ought to be honest with him really. I don't think it's fair on him not to tell him. Because if he has grown up with [husband] as a dad, then he's going always to think of him as a dad.'

Another couple put the point of view that not only did a child probably have a right to know of its origins, but that it was not a practicable proposition to keep it from the child in these days when AID is more common. The husband remarked, 'That's the only thing about telling your parents – I suppose later on in life you will have to [tell the child], whereas if you didn't –' His wife interrupted and finished the sentence for him' – you wouldn't have had to. But I wonder whether or not she is entitled to know anyway. It's all a bit deep for me! You can get into evasion – and, I mean, children aren't stupid either. Later on, things are talked about – "It took years before your mother had you." And they start to question it later on. And after all it's no great . . . there's nothing wrong with it at all – it's just a less known thing than adoption.'

As we saw in Chapter 5, some of the older couples had told their children, when they were older, of their AID origins. The parents had not initially planned to do this, but as the children grew up, unexpected and unpredictable factors had cropped up which made them change their mind. Questions were asked, behavioural problems arose, worries had to be dealt with, and these events made the parents conclude that it would now be best to bring AID out into the open. From the experience of these older couples, and the experience of the small number of young adults who were made aware of their AID conception, it would seem that the fears of most AID parents are unjustified. Parents *had* found it possible to explain AID origins to their older children, and these young people appeared to have been glad that they had, eventually, been told. None of them had found it a particularly traumatic experience; they had certainly been surprised, but some of that surprise was that their parents had felt the need to keep the matter such a close secret for so many years. Relationships had not been spoilt, indeed in some cases they had been enhanced. But parents, even when they are convinced that to tell their children is the right course of action, find it a daunting proposition. If parents are to be encouraged to tell their children they will need skilled help and support. Most importantly, legal changes are needed to give the child a non-discriminatory status in law. It is not realistic to expect parents to be open about AID if their children are then to be laid open to the stigma of illegitimacy. Parents will also need help in how best to set about the task of telling their child in practical terms. A 'script' is needed, similar to that available to adoptive parents. Adoptive parents have undoubtedly been helped by the suggestion that they should emphasise how they *chose* their particular adopted child. In some respects the situation of AID children is less disadvantageous than that of adopted children. The most difficult thing with which adopted children have to come to terms is that they were abandoned by their natural mother; that they were 'chosen' by a new mother may sometimes be an insufficient compensation. But an AID child is, above all else, a wanted child and has no experience of rejection. The young people who had been told of their AID conception felt pleased that their parents' desire for children had been so strong, and this made them, as individuals, feel particularly valued and loved.

We believe that the secrecy which surrounds AID is harmful and unnecessary. The confidentiality of the doctor's consulting room is essential, and the anonymity of the donor may be justified, but these concepts are different from the secrecy which leads to deception of close family members and of the child itself. Secrecy, whilst ostensibly being maintained for the sake of the child, is closely bound up with the concept of stigma, particularly the stigma of male infertility. In this situation secrecy is understandable but we believe it is counterproductive. Rather than encouraging the denial of the husband's infertility by assuring the couple that no-one need ever know, energies should be directed towards establishing an enlightened understanding of the concept of male infertility, through education and counselling. The stresses which infertile couples have to face should not be exacerbated by introducing further stress generated by the perceived need to hide the husband's infertility.

Some couples recognised the need for a changed public attitude. Wife 609 said, 'What I would like to see is people's attitudes towards it change, so that if it got found out, it doesn't matter – in the same way that attitudes to unmarried mothers have changed. You don't think twice about it now and yet a few years ago it was the most appalling thing; you were blighted for life. In a way that's how I'd feel now if it got out. It's not that we're ashamed of it, but people wouldn't understand. It alters your position in society.' In the next chapter we will look more closely at the concept of stigma which underlies the perceived need for secrecy, and at the stress which AID couples experience.

7

Stigma and Stress in AID

The distress suffered by many involuntarily childless couples is very great; the frustration of their desire to have children is extremely hurtful and a feeling of sadness often permeates their whole lives. Before scientific advances made possible the successful treatment of infertility, couples had no alternative but to accept infertility as an affliction they had to live with. In some respects the possibility of successful treatment increases the stress experienced by childless couples as they now may feel personally to blame if in their search for children they have not pursued every possibility. Many couples who were interviewed expressed the view that those who have children without difficulty could never understand the heartache experienced by childless couples in their efforts to have children. This distress is increased by the feeling which couples have that an inability to produce children is stigmatising; a condition which makes them feel inadequate personally and which they believe reduces their standing in the eyes of other people. One wife said, 'I think it's something that nobody is proud to admit, that they can't have children'. This feeling of stigma is attached to three areas of the AID couples' experience – to childlessness, to infertility in general, and to AID in particular.

125

STIGMA

The vast majority of couples who marry have children eventually and this is seen as the natural and normal thing to do. Getting married, by implication, means parenthood. When couples have been married for a while and have established a home they are expected, by friends and family alike, to have children. The group of friends with whom the couple mixed when they were younger and who married at about the same time will probably all have begun to have children and soon the childless couple may feel that they have been left behind and are the odd ones out, separated from the joint experiences of their friends. Their family too may begin to drop hints about it being time to think of having children and may start to exert subtle pressures on the young couple to conform to the normal pattern of events. When they still do not have children, couples may be made to feel selfish and different and pushed to the edges of family life. One husband said, 'My sister has children, [wife's] sister has children – it caused difficulties in the family not having children. As if it isn't bad enough not having them, we felt guilty.' So childless couples may come to feel that they are outsiders from life as it is usually experienced.

Couples who are physically able to have children may of course voluntarily decide to remain childless, and they too may have to endure the criticism of others because of their non-conforming behaviour. But couples who are involuntarily childless have the added burden of being infertile. It is not just a lack of children, but a personal disability or incapacity. They fear that others will consider them inadequate, to be teased or pitied if their disability is known. One husband who expressed his fear of being discredited in the eyes of other people said, 'I don't like people to think that it's me – that I'm any less of a person because I can't have a child'.

Couples also realise that AID, even though it is an impersonal clinical procedure, still transgresses the sexual norms of our society. The child is being conceived outside the marriage bond and this carries with it connotations of adultery and illegitimacy. By comparison, to obtain a child by adoption is seen as a much more acceptable procedure. Husband 1054 said 'People accept adoption, but when you say that it's AID there is some remark or stigma about it somehow'. Couples were not aware of any other couples who had had their children by AID and so they felt isolated and quite alone in their differentness. One wife said, 'When we were told about it we thought we were unique cases, because you don't realise how common it is'. The fact that AID was obtained through the medical profession helped to give it a more respectable status and to reassure couples that they were not behaving improperly. One husband said of his initial interview with the AID practitioner, 'I think that was the best thing. She actually said what had to be said and you had no doubts that what she said was a) what she believed and b) what she believed was right – which does help quite considerably.' However, some couples felt that because the NHS did not generally provide AID, and it could often only be obtained at expensive private clinics, this implied a certain 'shadiness' about the practice. One wife, who had transferred after initially beginning AID at a more expensive and luxuriously appointed Harley Street practice said, 'I felt – odd thing to say – but I felt it was all very much above board and medically *right* and accepted [here]. Whereas I felt I was able to have this treatment through London because we could pay; and that put a different complexion on it for me. It didn't feel – er – 'right', and I didn't want it to be that way. I wanted it to be natural and right as far as possible.'

Stigma is also seen to attach in different ways to three of the persons involved in the AID process; the husband, the wife and the child. Many of the couples expressed the

127

opinion that in the eyes of other people an inability to produce children would cast doubts on the husband's manhood. There was a generally held belief that having biological children was seen as a pre-requisite condition of full adult male status, and that by having children a man is made to feel 'a man'. Husband 235 said, 'I think the basic secrecy mainly is the hidden psychological thing against my masculinity'. His wife explained, 'They are men in every sense of the word, they still make love, they are masculine and everything, but because they can't produce children there is a stigma against them. I don't know what it is but a man is not a man unless he can produce children. It used to be *boys* – if a man could produce *boys* he was a man, but that has changed a bit now.' There appears to be a general lack of understanding about the concept of male infertility or sterility and this is confused with concepts of sexual impotency and lack of virility. One husband said, 'If you tell most people you are sterile they think you are not virile and you can be jibed about it and people just don't understand'. Frequently men confessed that their main fear was of ridicule; they did not expect other men to be intentionally cruel or hostile, but there would be an excess of teasing or 'taking the micky'. One husband explained, 'I know a chap at work – he and his wife they've been trying for two or three years, and the amount of stick that that chap takes – virtually everyone that knows him has offered to go and do the job for him. To be honest I would hate to think of the amount of teasing I would get if men at work found out. They would make my life hell – not nastily, but they would just go on and on.'

Although the AID procedure is a detached and clinical one, wives are conceiving a child outside the marriage bond. They are aware that their behaviour may meet with disapproval, particularly from the older generation, and that some may even consider their action to be tantamount to adultery. Some wives fear that if their AID were known their character might be compromised and people might question their moral standards. One wife explained that she would prefer to be able to talk about AID more freely but '. . . well, I always feel for instance that they're thinking, "Well, look at her, she's no good, it's not her husband's child"'. Others felt that they would be singled out for comment as being different because of AID. One wife had not told her new GP that her children had been conceived by AID. It was not that she doubted his confidentiality, 'No, just that he would know and that he would look at me as I came in and think "Oh yes, that's the one with the AID children"'.

Many parents expressed the fear that their child would suffer and might be hurt by unkind comments if others were aware of the child's different origins. They might be classed as different from other children, or they might be isolated and taunted at school because of their anonymous clinical origins and because 'their daddy is not their real daddy'. There was a general feeling that children can be particularly unrestrained and uninhibited in the way they behave towards each other. Husband 1039 said, 'People can be cruel. Children can be cruel if they knew.' And his wife added, 'It's not so much now that she's younger, but when she goes to school – you know – somebody has told somebody – and they turn round and say "He's not your dad". I can imagine children doing that because they can be really cruel if they want to.'

Individuals seem to be reluctant to acknowledge a fear of stigma on their own account. It was much more common for husbands to express a fear that their children might be stigmatised; relatively few were willing to admit that they themselves were afraid of being stigmatised. More often it was the wife who tended to express fears that her husband would be exposed to stigma. The interview with couple 334 illustrated this point well. While explaining why she thought secrecy about AID was necessary the wife said, 'I think for my husband's sake, because I think men get joked about . . .

129

Why I think about keeping it quiet is I don't know whether there could be any little jokes or things like that. I want them to think that they are his [children] as well as mine – not because I'm ashamed to say they weren't – but – I was thinking of [husband] more than me.' Her husband then continued, 'One of the main reasons I think of keeping it quiet is that there is nobody so cruel as children and I would not like them to have the 'micky' taken out of them by other children for something they could not really understand'.

Perhaps the most striking observation was that the stigma attached to male infertility was perceived as being much greater than that attached to female infertility. It was common for the wives to pretend that it was they and not their husbands who were the infertile partner. One reason for this may simply have been that male infertility is known to be more difficult to treat, and so it would have been more difficult to present the child as a natural conception of the marriage following treatment. Wife 703 explained that this was why she had pretended she was the infertile partner. '. . . Because I read everything I could on the matter and I felt there was nothing, well there was very little that could be done for male infertility. There were lots of things that could be done for female infertility. Once you cast a doubt on male infertility there's very little could be done. And if I'd cast a doubt in their minds . . .' But in the majority of cases the wife had taken on the infertile role because it was felt that female infertility was more sympathetically viewed and more acceptable to the general public; people did not look down on a woman who could not have a baby in the same way that they would look down on an infertile man. Wife 906 explained this viewpoint. 'I think it's different attitudes. I mean, if women say they are having difficulty getting pregnant everybody is usually sorry for them and they can be helped. But if a man says he can't have children a lot of men seem to think this is something funny and they are not proper men because of it.' Not only was this differential stigma made explicit in the comments of the couples who were interviewed, it was also often implicit in their behaviour during the interview. In a number of cases the husband displayed a greater degree of anxiety at the beginning of the interview than did his wife. This was sometimes signalled by a reluctance to answer questions and to enter into the discussion, and sometimes by a certain amount of hostility and aggression in response to questions. In almost all cases this initial display of anxiety eased, and the husband became more relaxed and confident as the interview progressed. Presumably this greater self-confidence was generated by the realisation that the interviewers understood the husband's feelings (as far as they were able) and did not see male infertility as a discrediting feature.

One reason for the greater stigma which is attached to male infertility may well be that it is not well understood. Compared with female infertility the causes and mechanisms of male infertility remain relatively obscure and so successful treatment is rarely possible. Diseases which are mysterious and relatively uncontrollable, for example mental illness or cancer, are more threatening and feared than other better understood conditions. This threatening quality is thought to contribute to the degree of stigma which commonly attaches to these mysterious conditions. Thus, the stigma attached to male infertility may be increased because the cause of much male infertility remains obscure and effective treatment is unavailable. This threatening quality of male infertility may be one reason why so many doctors, most of whom are male, appear to find it extremely difficult to break the news of infertility to their male patients. An unexpected finding of the research interviews was the dissatisfaction of many husbands about the way that they had been told that they were infertile. Nine husbands were spontaneously critical of the way their doctors had told them of the results of their semen analysis. In four other cases the doctor had evaded the issue by giving the wife the results of the tests in

131

the absence of her husband. It may well be that the male doctor's interaction with his infertile patient was itself being affected by the threatening quality of male infertility. Another researcher has observed, 'Most frequently the diagnosis is given with embarrassment and the doctor's fugitive attitude signifies both the gravity of the situation and the impossibility of discussing it. Now begins a phase of solitude and fear of social rejection.' (Czyba and Chevret, 1979)

But the differential stigma of male and female infertility also appears to be related to sex-role stereotypes. The male stereotype emerges in instrumental forms of behaviour; that is men are expected to take the initiative, to be strong, to achieve, to strive and to succeed. In the event of failure a man must bounce back and try again until he does succeed, and it is considered weakness and a defect of character to become emotional over a hurt or a failure. Thus a man's infertility conflicts with his instrumental role; it reduces him to a state of powerlessness and so is incompatible with his socialised concept of appropriate masculine behaviour. He feels impotent in a social context even if he is not physically sexually impotent, and so concepts of infertility, impotency and lack of masculinity become confused. The man fears that he has become less than a man and will be discredited in the eyes of other people. Conversely, the female stereotype emerges in expressive forms of behaviour. Ideally she cares for others, she nurtures them enabling them to fulfil their potential. It is socially acceptable for her to be dependent on others, to be weak and to express emotions of grief over hurt or failure. This means that a woman's response to infertility can be consistent with her expressive role and she can react to it in ways which do not conflict with appropriate feminine behaviour. This helps to explain why it is so common for women to pretend that they are the infertile partner rather than their husband. A woman feels her role is to care for others, therefore it is appropriate for her to care for her husband by taking his infertility upon herself; and a woman can do this without compromising her feminine image. Furthermore, wives have been socialised to desire a 'masculine' husband and so may have their own interest in preserving his masculine image. One husband said, 'I suppose there will come a time when AID is accepted as just another form of treatment. I suppose it's something to do with male chauvinism really. If it's the woman, we half expect the woman to say to the world, well, I needed this treatment or that. But I would find it very difficult to tell the world I had a low sperm count.' His wife, who had taken on the role of infertile partner, interjected, 'I think that's very understandable. If I was the bloke I'd feel the same.' Her husband continued, 'Although if you could introduce me to other guys – and I've met them in the National Association for the Childless – you realise that they're ordinary chaps'.

STRESS

In the planning of this study it was hypothesised that parents of AID children would find it stressful to maintain the secret of AID, and that this stress would increase as the child grew older. Interviews with the parents of young AID children revealed that they had indeed experienced stressful situations but not in the way that had been hypothesised. The stress experienced was greater in the period before AID was begun and during the AID session than it was after a child had been conceived. The states of childlessness and infertility are also stressful, and although AID carries its own stress, it may reduce the stress of infertility and childlessness. Wife 241 said 'It's really solved our problem. We wanted children and we would have done anything to have had children, and that came along and solved our problem.'

It has already been noted that the desire for children was seen by couples as a natural, instinctive desire. Couples had married in the expectation of having children in the natural run of events; the only decision to be made was when to have them and they were taken by surprise when a pregnancy failed to materialise. Having children had been a central assumption of their expectations of married life and when this assumption proved incorrect it was as though the bottom had dropped out of their world. Wife 895 said to her husband "It was a shock that you haven't really got over, wasn't it? Both of us really. Because you don't expect it – not yourself – not really.' Children were seen as a physical demonstration or manifestation of a couple's marital relationship, something that they had created together as a result of their own union. Couples also saw children as a gift from one to the other, something they could give to each other. Several husbands admitted that when they first knew of their infertility they had offered their wife a divorce because they could not give her children. Wife 878 described her reaction to her husband's offer of divorce. 'I said "What for? I want children but I want *your* children not anybody else's." And the way it came out, I class them as [husband's].'

It was also taken for granted that having children is a superior way of life to remaining childless. Couples believed that children enhanced a marriage relationship and that by not having children they were missing out. Husband 1062 explained how he and his wife had felt deprived by not having children. 'I don't know, it seemed to mar our holidays by seeing other people with children. We used to love children. Every time we'd go off for the day, we used to send down the village and see if the X children would come with us, and we used to take children off with us to make our day. I said to [wife] there's something wrong with our life, you're going to see your sisters grow up and have children and we're going to be left without any children.'

A wife, though not infertile herself, nevertheless shares the experience of childlessness with her husband. Wife 875 said, 'When I was holding someone's baby, or even looking after little children, I'd play with them, and do all sorts, and it was lovely. But when I went home, I could have climbed up the wall, because there was so much love inside of me that I couldn't give.'

One husband (334), describing his reactions on being told that he was infertile, said 'Anyone who's never actually been told, you can't tell them the feeling that you get. It's like being hit with a sledge-hammer. When the doctor had told me he sat me down and dashed out of the room to get me a drink. I think I was going to fall over then and there – I never felt so ill in my life. It took me a long while to get over it.' Some husbands stressed the feelings of inadequacy. One husband said, 'I felt, well, a bit useless at the time'. Another said '. . . That's the worst, you feel incomplete'. Some men tried to compensate for these feelings by deciding to concentrate on trying to achieve success and promotion at work.

A few of the more thoughtful husbands were disturbed that the continuity of reproduction, of ancestor and descendant, had for them come to an end. Husband 902 said, 'My major hang-up really was based on this rather metaphysical notion of genetic immortality. What depressed me most of all, and overwhelmed me mentally, was this idea that at this point my genetic channel stops. That's the end. And that was the most chilling thing that I had to take on board.' Children were seen as continuity with the future, a way in which parents would live on in future generations, almost as a means of achieving immortality. These husbands saw the finality of their infertility as a kind of genetic death, and AID did not cure this problem for them. AID did not cure their infertility, it merely circumvented it. Later in the interview husband 902 explained how he felt about this. 'I must be frank and say that I still wish that I could father a child, this is still a faint note of sadness to me. But it doesn't intrude into my relationship with these two children, because I can really put my hand on my heart and say that if I could turn the clock back now – I wouldn't really change them, that really doesn't make any difference at all. But I think if

135

somebody suddenly discovered that fertility had come back I would want to try most vigorously to have a child. But I think that's a very primaeval and natural response.'

Despite the fact that AID does not cure a man's infertility it does cure his childlessness, and if AID is kept secret it allows him to deny his infertility by pretending that his wife has conceived naturally. Because of this, AID may be viewed very positively by the man as a hopeful solution, and this perception tends to balance the negative connotations which AID may have for him. Therefore many couples see AID as a stress reducer and a solution to their problems. Nevertheless AID of itself and the secrecy which surrounds it does produce other stresses. In addition the conflict of roles between the husband and the donor produces a state of dissonance which must be resolved in some way.

During the process of AID and up to the birth of the child, couples may be anxious about how they will react to the child. Wife 609 said, 'You are going completely in the dark. You don't know how your own feelings will be towards the child when it is born. You don't know what your husband's feelings will be about it. After the baby is born you know it is the right thing; there's no problem.' All couples who are about to have a baby are concerned that the baby will be normal and not malformed in any way but these fears take on an added significance when the child has been conceived by AID. As we have just seen, couples are slightly anxious about how they will feel about a normal baby – will they be able to accept it as truly their own, or will it be seen as some sort of imposter? But couples were particularly worried about what their reaction would be should the child be handicapped. One wife (1009) said, 'I felt during the pregnancy that [if the baby was handicapped] it would be just too much. From my husband's point of view I felt it [AID] was enough anyway – and for it to have been a handicapped baby would have been too much.'

Wives were aware that their husbands might feel jealous.

One older wife (849) said, 'Naturally at that age you are still very sentimental – and the fact that it wouldn't be his – would he think, Oh I've got another man's child in me?' Understandably, most couples were concerned about the qualities of the anonymous donor. They had needed reassurance about his health and general character, but above all they were anxious that he should look like the husband. If AID were to be kept secret, the child must be of an appearance compatible with that of his parents. One husband joked, 'It wouldn't be a good idea to have a baby with ginger hair, and a ginger-haired milkman and a husband who's got dark hair would it?'. Colouring of hair and eyes was seen as particularly important if the child's appearance was to be compatible and give rise to no questioning remarks. This worry grew as the time of the birth drew nearer, and some wives confessed to spending the last few days of pregnancy worrying if the child might have red hair, or a long nose, or even, by some awful mistake, be of a different racial origin.

But the main cause of stress during the AID process was undoubtedly the stress of failure to conceive. After each visit for AID the couple were hoping that the insemination would have been successful and that the wife would become pregnant; when their hopes were dashed they became very disappointed. If this happened on repeated occasions the cycle of hope and continual disappointment became extremely stressful and the couple's resilience became worn down. Couples who spontaneously emphasised the stressful nature of the period of inseminations had needed on average twice as many inseminations before a conception occurred as couples who did not complain of stress; among the seventeen couples who spontaneously emphasised the stressful nature of the period of inseminations, the average number of inseminations per live birth was twelve. Among the remainder of couples who did not complain of stress during AID, the average number of inseminations per live birth was six. In addition, couples who were at the time of the interview attending for further AID in order to have another child, were also more likely spontaneously to complain of stress; seven of the eleven presently attending couples did so. One wife (1004) said 'I was craving for a child. We would have paid with money for a child at that time. Now we look back and say we couldn't possibly have done that, but at the time we did consider.' Her husband put in, 'It's fine telling someone to take their temperature every morning; it's another thing doing it'. His wife continued, 'The disappointment when your period starts, it's so intense because for a full two weeks you are hoping, and then you go down so quickly. And of course then you have to wait for another two weeks before you can have another chance. When you sum it up you seem to have twelve chances a year; twelve chances out of three-hundred and sixty-five seems so minimal.'

Daily temperature recording was frequently identified as a cause of stress. Wife 664 said, 'One of the most depressing things I find is that you can never, ever put it out of your mind and forget about it, because every morning you have to take your temperature – it's always with you whether you're on holiday, or Christmas, or whatever'. This woman had learnt to 'read' her temperature chart and to know whether it was likely she had conceived or not. She pointed out that 'It was particularly depressing to have to keep on taking your temperature when you knew that this month was a failure again and you were going to menstruate'. This stress associated with failure to conceive, of itself further compounds the chances of failure as it is known that psychological stress can cause disturbances of the menstrual cycle which will reduce the likelihood of conception.

The perceived need for secrecy, and the resulting isolation which these couples experienced also increases stress. Many couples who tell no-one at all outside the medical profession, have no confidante with whom they can share their disappointments and fears. The crisis-meeting resources of the family are not mobilised because the crisis is hidden and the couple have to cope alone as best they can. These couples often felt an overwhelming need to be able to share their problems. Wife 875 said, 'It took us eighteen months – maybe longer than that – and we were in ever such a turmoil. There was nobody we could go to . . . I think that's what you miss more than anything – just to go somewhere, sit down, and cry your heart out and say "What can I do?"'.

When a couple have a child following AID there is confusion about the paternity of that child. Although the husband may believe, and we may agree with him, that he is to all intents and purposes the father of that child, he is not the genetic father; that role was taken on by the anonymous semen donor. This confusion is illustrated by the way many husbands, whilst considering themselves to be the father of the child, sometimes refer to the donor as 'the real father'. Though we may accept that the husband is the child's father in practical nurturing terms and the donor a much more shadowy, perhaps less important figure, it still cannot be denied that both men are father to the child; one a genetic father and one a nurturing father. This contradiction produces a state of dissonance which is stressful to the couple and which they feel a need to resolve. The interviews revealed that couples reduced this state of dissonance by two main strategies: a defensive denial of the role of the donor, and a rational minimisation of the role of the donor.

To acknowledge the fact that the child was the genetic procreation of another man was for most men a hurtful and traumatic experience. Some men found it so potentially hurtful that they evaded this acknowledgement by a defensive, psychological denial that the donor was in fact the genitor of the child. Denial of the role of the donor was not limited to the husbands; wives and other relatives who had been told about AID often joined in this denial. Various factors facilitated denial. It was claimed that the true identity of the genitor was ambiguous; no-one could know for certain that it was not the husband. The husband was continuing to have sexual intercourse with his wife and although previous semen specimens had been shown to be azoospermic, no-one could be absolutely certain that this was so for all ejaculations. One wife explained how her sister-in-law had remarked after the baby was born, 'You went up there for eighteen months [for AID] and didn't get anywhere, so who's to say that [husband] didn't come up trumps in the end? It's just as likely.' This claimed ambiguity is strongly reinforced by the advice given in the information leaflet for couples prepared by the Royal College of Obstetricians and Gynaecologists (1979). In the section dealing with the legitimacy of the baby, couples are advised 'Strictly speaking, a baby conceived by AID should be registered as "father unknown" . . . However, babies born within a marriage are presumed to be legitimate and provided you do not abstain from intercourse during the period in which AID was carried out *there can be no certainty that any child conceived is not your husband's'* [emphasis added].

Denial is also aided by the clinical nature of the procedure; no man is involved but merely a doctor in a white coat and the usual medical instruments. Wife 902 said 'While I was pregnant it never entered my head – or I never really thought that it wasn't [husband's] child, because it's so impersonal'. This attitude is sometimes taken a step further by asserting that donor insemination is merely medical treatment. Husband 235 said 'I can't even picture it as a donor to be honest with you. It's just a medical treatment the same as an operation is a medical treatment. It's the medical treatment for the medical problem, just the same as you take aspirins for a headache. I've got infertility, so you take "aid" to solve the problem that way.' If the wife is also subfertile to some degree and also requires medication, this view is reinforced.

Later on, as the baby grows, he or she may bear some physical resemblance to the husband, and frequently may begin to take on some of his mannerisms and so denial is again made easier. Husband 975 said, 'I keep thinking, well perhaps he *is* mine. I keep looking, I keep looking at the shape of my hands, and this, and that, and the other. His blood group is the same.' Friends and relatives all treat the husband as the proud father of the new baby and this social confirmation of fatherhood facilitates denial, as does the legal confirmation of paternity when the husband registers the child as his own. One widow who was interviewed illustrated this point: 'Don't you think very often when you have an AID child, the husband is rather pleased to feel that he *has* been able to have a child, and goes and registers it rather proudly? I'm sure my husband did. I'm sure because they've had difficulty in having it, therefore when they do have one everyone says "Oh, isn't that nice", and they are rather specially pleased to go and register it as theirs.' Even the availability of words in the English language facilitates denial. There is no separate English equivalent for pater and genitor, and the husband is, to all intents and purposes, the child's father. Similarly the term 'son' describes the position of the child in relation to genitor and pater. Couple 235 said 'He *is* our son; there's no other way to describe him'.

Some couples do not deny the role of the donor and are able to acknowledge that the donor is the genitor. But for the child to have two fathers is still a social contradiction and this is resolved by rationally concluding that the role of genitor is unimportant compared with that of the nurturing father. Parents stress the social reality of the situation and minimise the genetic reality. Temporally the role of any genitor is brief; once the ovum is fertilised his role is complete. The pre-natal carrying role of the mother is more lengthy, therefore couples stress her role in the procreation of the child over that of the donor. Husband 780 said, 'As far as we were concerned the baby would be ninety-nine per cent [wife] and one per cent lent from someone'. His wife developed this idea, 'That's always been your [husband's] view, hasn't it? Which in a sense is true enough. After all, the woman carries it for nine months so it's got to be more of the woman, hasn't it, if you look at it that way?' The brief role of the genitor is also contrasted to the lengthy role of the nurturing father. Wife 878 said, 'Because apart from not putting the sperm there, [husband] has been there the whole time'.

The evidence about the relative influence of hereditary and environmental factors in development, the nature–nurture debate, is used to stress the importance of the role of the nurturing father over that of the genetic father. Husband 004 commented, 'I feel very strongly that we do impart more than the major part of the child, rather than it being hereditary factors. I believe the parents create the child rather than the child creating itself.' The anonymity of the donor is also used to diminish the importance of his role; there is no identifiable third party who may one day turn up on the doorstep and cause embarrassment or make demands.

Paradoxically, a few husbands attempted to resolve the conflict with the donor, not by diminishing his role but by stressing the superior characteristics (usually health and intelligence) of the donor; their child would benefit from these superior characteristics. Husband 975 said, 'Really and truly I have no desire, I know this sounds ridiculous but, I have no desire to reproduce *myself*. Because I've got my faults, and I'm not all that brainy, and perhaps [child] will be a darned sight brainier than I am.' These husbands were proud of their clever children, and found some compensation for their own infertility in the achievements and reflected glory of their children. Another husband (906) said, 'Sometimes, when [child] comes out with one of her astonishing things she does, I sometimes think to myself, 'Who *was* that guy – because she's so clever?'

There is no doubt that besides a defensive denial of the role of the donor, there is also an honest 'forgetting' for most of the time, as AID treatment recedes into the past and ceases to be a current problem. This is functional in reducing stress and facilitating normal family relationships, and a similar phenomenon occurs in adoptive families. Nevertheless, this forgetting is not complete and sometimes events occur or things are said which bring AID back to mind again. Couple 1054 had told all their family and the husband said, 'I think everyone basically forgets'. But his wife replied, 'I don't think it's a thing that's constantly on people's minds – but it's not a thing that's totally forgotten either. People forget about it, it goes to the back of their minds, but its never totally forgotten.'

The implications which the AID procedure holds for family relationships are many and complex, and we believe that couples should be assisted to examine these implications openly and honestly rather than burying them. Denial of the true state of affairs and the pretence of normality is reminiscent of the practice followed by adoptive parents several decades ago. Since then it has been established that the emotional energy spent on denial and concealment is better expended in facing and resolving the inherent problems. But parents need help to do this. Almost all the couples interviewed expressed a need for counselling; they would have valued an opportunity to explore their hopes and fears and disappointments with a skilled and sympathetic counsellor. Many couples found that talking about their experiences during the research interview gave them a rare, sometimes a unique opportunity to examine and talk out their fears and hopes, and that this was helpful and therapeutic. We believe that the team providing an AID service should always include a trained counsellor who can help the couple to reach a well-founded decision about AID, and to whom the couples can have free access during the AID process, and after the birth of the child.

Part III

THE SOCIAL IMPLICATIONS OF ARTIFICIAL REPRODUCTION

8

Artificial Reproduction and Society

Throughout this book the assumption has been made that those special relationships which encompass the 'family' are extremely important for the individual members who make up that family. The quality of these relationships is hard to define but we have determined that it contains a mixture of perceived rights, duties and obligations resting on a base of trusting expectations we hold about each other. But the concept of the family has a wider application than merely the psychological well-being of the individuals concerned. The ideal family situation is not only a place of personal security and support but also an integral part of the whole social system in which the family exists. Indeed, some would argue that the family is the basic unit of social organisation and that the form of the society in which we live is built upon an ever more complex interaction of social relationships which start in the family.

It is not difficult to see why this should be so. The inculcation of values held by the wider society is only possible if undertaken early in life and the socialisation process we have all experienced owes much to the control exercised within

147

our family circle. The family is an organisation whose function is to provide mutual support for its members and nurturance for the children it produces or acquires. Being the group with which a newborn infant first comes into prolonged contact it shapes the development and outlook of the infant and child, imparting to him or her (at times differentially) a pattern of appropriate behaviour, especially that associated with the norms that define kinship. In so doing, the family as a small social organisation helps to determine the values, customs, morals, social practices and general outlook of the child. To be sure these may all be 'tested' at adolescence or at other times, but even if some contrary reactions do occur, basic values are imparted and provide effective, persisting bonds between family members. The transmission of societal values from one generation to the next requires a socialisation process that is effective and enduring. How we relate to other people of the same or opposite sex, our views about openness and privacy, and the feelings of shame or guilt which we sometimes experience are all heavily dependent upon our early discipline and upon our imitation of those who surrounded us at a formative stage in our development.

The intellectual or rational views we hold about our family relationships can be fully appreciated only if the emotions which surround this rationality are acknowledged. The expectations we hold about trust, honesty and integrity are as much emotional as rational, but without them no society could endure. Without such values as trust and truthfulness the predictability of behaviour required in all societies would be absent. If statements were randomly truthful or false there could be no planning, no expectation that what is promised will be carried out, no law and no certainty. In short, no order but anarchy and chaos. We expect the bus to travel to the destination marked on the front and the cost of petrol to be that shown on the meter. We believe people when they provide information from an authoritative position and we feel angry and frustrated if any of this trust is shown to be misplaced.

SOCIAL SYSTEMS

The claim that the family is the basic unit of the social system within which we all live raises questions about the nature of the social system itself. The word 'social' indicates that the system being described is concerned with human organisation in which large numbers of individuals are participating, but the definition of a 'system' is not so straightforward. The most obvious feature of a system is that it consists of a number of interdependent parts making up a differentiated whole. This implies that what happens in one part of the system has effects on the other parts. Another feature is that while the system is dependent upon its constituent parts, the totality of the parts creates a dimension of its own. For example, a bicycle is a mechanical system made up of a number of parts; the wheels, handlebars, gears, chain, saddle and frame. Without these, the bicycle will not function as a bicycle is intended to function. But over and above the sum of the parts, it is the interdependence of the parts that allow us to view their particular relationship as a bicycle. There are many different systems. For example, within the human body there are systems composed of different organs which work together in an interdependent way. The parts of the systems described can be viewed at a micro-level (constituent parts of the saddle, cells which make up the organs, etc.) or at the more broadly based macro-level, but the principle of interdependence remains. So it is with social systems.

The parts of a social system are usually viewed in two ways. On first visiting another country with a social system different from one's own, the careful observer who wishes to know the distinctive nature of that society will attempt to discover the principles underlying the basic units which

149

make up that society. For example, one will want to know something about the legal, political, educational, religious, economic and kinship structures of the society in question. If one has the time one will want to discover how each of these structures are controlled and what rules govern their maintenance or breakdown. Each structure contains other more closely-defined elements which serve to maintain the structure as a whole. Thus, the legal structure may contain a number of parts including the creation of laws, the availability of courts and the presence of a police force, etc.; similarly, the kinship system may contain rules relating to marriage, reproduction and the care of aged relatives. As the bicycle has its wheels and spokes, so society has its legal system and policing policy. At each level the parts contribute to the whole. Sociologists call constituent parts of a social system its 'social institutions'. Each social institution has come about over a period of time in an attempt to ensure the smooth running and orderly nature of that society. The same rules apply as for the mechanical and organic systems. The whole comprises a set of interdependent but differentiated parts; the social institutions described are interdependent and a change in one will effect changes in the others.

Such social institutions are themselves made up of parts; thus a legal system is made up of laws, sometimes highly systematised as in a legal code such as the Code Napoleon. Kinship likewise is made up of parts, that is to say the rules defining descent. These institutional parts are social norms of one kind or another. It is a great mistake to suppose that human society is a horde or collectivity of people. Human society is very much the normative order which enables individuals to live in relationships one with another. It is this institutionalised behaviour by a multiplicity of people, often differing from each other in many ways, which enables society to cohere. How else may people identify with their village or town, their region or nation? Of course the cohesion is tested and strained at times and there may even be social disruption, but when this happens people tend to take sides, which while displaying differences also reflects an underlying acceptance of the framework within which disagreement is expressed.

A second way that the interested observer could describe a society is by reference to how groups of people interact with each other within social organisations. Whereas the social institution defines normative behaviour in a given situation, the study of social organisations describes how people interact with each other in terms of the positions they occupy and the roles they play. In some respects we have moved from the more abstract level of determining the rules by which a social system operates to the more concrete level of the organisational structures within which people interact, but they are both equally important. The normative behaviour which is expected within the institution of law has to be matched by the positions held and the roles played by those performing within it. The judge, a juror, a policeman or a defendant are expected to behave in an expected manner within the legal system in accordance with the positions they occupy and the roles they play. It is the same for the institution of kinship in relation to the social organisation we call the family, as we showed in Chapter 3.

We can now see that both norms and roles are important guidelines for behaviour. To have to reflect before every behavioural act would be an impossible burden. So much has to be taken for granted and as behaviour is largely repetitive and stereotyped, regularities are provided in conformity to norms and expected actions in a given situation, based on mutually understood role-playing behaviour. Such social behaviour results from *mutual* expectations, for most people most of the time are predictable in their behaviour; to be otherwise would be deeply disturbing not just for family life but for social life in general. It is thus that we can speak of institutionalised behaviour, learned through a process of socialisation and dictated by conscience and experience.

Yet another aspect of human interaction is the degree to which the norms and roles which define behaviour are formally or informally stated. Clearly some norms are presented as formal legal requirements and are defined in law. The legal responsibilities of parents for their children, and the rights of married partners in relation to each other are examples. But clearly there are behavioural expectations that do not have the force of formal prescription in legal terms but which are equally important to family life despite their informal nature. Thus it may be well known in the family that father must not be disturbed after Sunday lunch, or that grandmother must be visited or telephoned at least once each week. Sometimes it is difficult to disentangle the formal from the informal; indeed, so much is unspoken yet well understood. It is not just the use of a private language of special words or gestures peculiar to a family that we refer to, but a whole range of cues, sometimes perhaps imperceptible to non-members who visit the family. Herein lies the intimacy of family life which shares its secrets, its knowledge about the characters of family members, their dispositions, weaknesses, desires and hopes. As we said in Chapter 3 it is this kind of family life which engenders trust and reliance, love and caring, and which in turn provides the security so necessary to the development of the child. It is not, we hasten to add, all smooth sailing, far from it. The relationships may be forged at times in the heat of revolt, they may be tempered by the chill of distress, loss, bereavement, jealousy, irritation or plain dislike. It would be foolish to believe that there can be sound child development without any emotional disturbance or clash of wills. But the essential characteristic is that the institutionalised behaviour, and the roles people play whether in a formal or informal capacity, are based upon the assumption of predictability and consistency, truthfulness and trust. There is a fundamental trust that brothers, sisters, mother, father, aunts and uncles really are the people one is told they are. If this were otherwise, all manner of social difficulties would ensue.

From this brief introduction to the study of the social system it can be seen that the family unit occupies a peculiarly important place in the maintenance of society. The family in one form or another is ubiquitous, it is where the values that lie behind all other social institutions and all forms of social organisation are first generated and transmitted. The family is the place where intimate relationships most frequently occur, where the objectivity of relationships and their subjective feelings are first combined and where the dual importance of formal and informal rules of conduct are presented. Other social institutions also play an important part in the development of the social being, education and religion are obvious examples, but the intimate relationships of the family circle come earlier and are likely to set the scene for much that follows. The family is the place where the first experience of normative order is introduced based upon values associated with the predictability of behaviour.

CONTROL OF REPRODUCTION

Yet the most important function of the family lies in its role as the reproductive unit of society, enabling social life to be perpetuated. Before socialisation of the child can take place, the child must be conceived and born; and before the child can be conceived, it is normal practice for a male and a female to take part in an act of sexual intercourse. It is useful, at this point, to reiterate something that was mentioned briefly in Chapter 1. No society anywhere, at any time, has left the issue of reproduction entirely to chance. Particular means for both restricting and encouraging either an increase, or a decrease, in the number of births have been devised but the control has not been a direct one. An indirect control is much easier to maintain once it is accepted that reproduction results from the act of sexual intercourse. In simple terms rules directly controlling reproduction are relatively un-enforceable but rules governing who should have sexual intercourse with whom are much easier to contain within a formal legal framework. Examples of this indirect control are the incest taboo, the age of consent, the ideology of pre-marital chastity, and the prohibition of marriage within certain degrees of relatedness.

The repugnance for sexual intercourse between close relatives is often claimed to be due to the wish to avoid congenital malformations and disease associated with 'in-breeding' but this repugnance remains even when reproduction is unlikely to occur owing to the use of modern contraceptives. Again, part of the repugnance for sexual intercourse between close relatives may stem from the hazards of giving birth to a child for whom there is both genetic and social confusion; a child who is both son and grandson at once must surely create difficulties in role-playing within the kinship group, quite apart from any question of the consequence of in-breeding. Nevertheless, sexual intercourse between close adult relatives is seen as socially undesirable even where the likelihood of pregnancy is virtually non-existent. The reasons for this may appear to be historical and based on outcomes which are no longer the probability they once were.

The prohibition of marriage among close relatives is one of the most interesting indirect means of controlling repro-duction, for despite its close association with the incest taboo, it contains two assumptions that have debatable rele-vance in the last decades of the twentieth century. The first assumption is that sexual intercourse should normally take place only within marriage and the second is that sexual intercourse is for the purpose of reproduction. When these two assumptions could be realistically held for the majority of the population, the indirect control of reproduction was to this extent complete.

It may be argued that the arrival of contraception has

effectively separated sexual intercourse from reproduction and this means that marriage no longer has the social force it once had in restricting sexual intercourse. The rise in the numbers of cohabiting young couples demonstrates this change, but it should be noted that the majority of such couples marry when pregnancy is either desired or experienced. The point being emphasised here is that the laws, customs and conventions surrounding family life have evolved over centuries when it was not possible to separate the sexual act from reproduction as has become the experience of more recent times. The basic issue of who has sexual intercourse with whom remains at the centre of family organisation, so much so that the formal rules relating to consummation of marriage and to adultery are still taken very seriously.

It is worthwhile asking why the state should be so concerned for family life and the role the family plays in the procreation and welfare of children. It is a simple question and the answer perhaps is obvious; children are tomorrow's citizens, and the entire society through the medium of the state, is interested in their welfare. This is because the basic values of our society are instilled in infancy, such values as respect for other people, truthfulness and honesty, a sense of duty and obligation to others and at best a creative awareness and activity which enables society not just to survive but to develop. Society, therefore, must view with disquiet anything that impairs the contribution the family makes to the society of which it is an essential part. In the absence of alternative structures society still has to rely on the family for the performance of significant tasks.

SECRECY AND ARTIFICIAL REPRODUCTION

It was noted in Chapter 1 that where artificial reproduction is confined to the fertilisation of the wife's ovum by her husband's sperm there are no direct social implications that require consideration. There may be indirect implications where the procedures developed for this purpose are used in other situations, but the procedures of AIH and of external human fertilisation (EHF) followed by embryo replacement do not, of themselves, raise issues of social significance. But uncertainty does arise when the use of a donor providing semen (AID) or ova (EHF and embryo transfer) is involved. The relationship of the child to his or her genetic and social parents becomes blurred and the rights, duties and obligations the definition of these relationships requires become socially confused. This confusion relates to both the formal and informal rules on which social interaction is based. We have already noted that one of the most important expectations governing human behaviour is the predictability of that behaviour and therefore of the trust which underpins it, and that the inculcation of such values as honesty, on which any trusting relationship must depend, begins in the socialising functions of family life.

If the child resulting from artificial reproduction techniques using donated semen, ova (or both) is deliberately kept in ignorance of his or her genetic origins by the people responsible for inculcating the values of honesty and trust, then the basis of social life in which all human beings have rights, is being undermined at its very source. The almost universal desire of parents using AID to keep the matter secret from the child and other kin indicates an area of conflict between social and personal needs. The adoption experience has shown that ways can be found to avoid potential conflict of this sort but the similarity in the circumstances surrounding adoption and artificial reproduction is limited. Nevertheless, adoption procedures demonstrate the only precedent that even remotely approaches the social implications of artificial reproduction.

One of the principal features of adoption is the recognition that the child has a right to know who his or her genetic

156

parents are if this knowledge is available. The social significance attached to this right is demonstrated by its formal
acknowledgement within the legal institutional framework
of our society. The adoption experience also demonstrates
that a person can become part of a kinship network in both
formal and informal ways. This suggests that the need for
secrecy in order to protect the AID child is of doubtful
significance. If the secrecy which surrounds most cases of
AID were to be removed, many of the questions raised about
its propriety would also be removed. AID parents would no
longer need to take part in a charade when making regular
visits to the clinic; they would no longer have to act deceit-
fully with those for whom they usually have deep affection.
But if parents are to be encouraged to tell AID children about
their different genetic origin they will need to have certain
minimum information about the donor which they can pass
on to the child. It should be made possible for the parents to
know something of the personality, appearance, interests
and activities of the donor and also to have access to a
medical history in the event of the child's illness. To give
such information would help to satisfy the need of the child
for some knowledge of its forebears. This does not mean
that the identity of the donor need be revealed or that the
guarantee of anonymity which the practitioner gives to the
donor need be broken. The withholding of information
about the donor from the recipient family by the doctor is an
important issue and should not be confused with the confidentiality which patients expect when seeking the advice
and help of a member of the medical profession.

THE PROVISION OF ARTIFICIAL REPRODUCTION

There is little doubt that when considering artificial reproduction the relationship between those seeking and those
providing the desired service is an important one. The exami-

nation of AID provision as a means of identifying the provision issues relating to all forms of artificial reproduction is appropriate. Artificial insemination is the most common procedure and involves most of the issues relating to the social, legal and ethical considerations that require resolution for all forms of artificial reproduction. The providers of artificial reproduction services and the researchers who support them usually belong to a profession which exerts some control over their work. Occasionally, but exceptionally at the present time, people outside the structure of a professional group provide this service; indeed a 'do-it-yourself' service in relation to artificial insemination has been undertaken. But the bulk of AIH and AID services and all the provision of external human fertilisation is being undertaken by professional people at the present time. It is therefore important to determine what is meant by 'professionalism' and what constitutes a professional relationship.

This subject is by no means an easy one to handle for professionalism is a somewhat elusive concept. We may recognise a profession to be a special kind of occupation, whose members are permitted certain privileges; it is an elite group. Over time it acquires a degree of autonomy which is recognised as legitimate, but only so long as it manifestly makes a valuable contribution to society generally. All professions require a period of training and the acquisition by their members of appropriate qualifications after examination. They usually have organisations which set standards and revise curricula, support research and publish its findings. Frequently, the professions are self-recruiting and also take on the task of disciplining those members who infringe the rules or fall short of minimum standards of behaviour and in their performance of their duties. But essential to a profession is a code of honour or practice. Thus there are rules, both formal and informal, regulating the behaviour of professional people. In return for privileges and a high social status members of the professions are expected to conform to certain 'professional' rules of conduct. These almost always refer in one form or another to the values of trust-worthiness, discretion and confidentiality. A professional person providing a service to a client or patient is expected to provide an expertise which is of benefit to that client or patient and therefore, more indirectly, of benefit to society generally. A member of the public consulting a solicitor expects to receive sound legal advice; a medical consultation leads to the expectation that any medical intervention will be for the patient's benefit. The key word in all these examples is 'expectation', a word we have used before when describing the roles of family members. Like the role of 'mother' or 'wife', there are also professional roles which are based on the mutual expectations of those playing that role or reacting to it. If professional expectations are not met or if the benefits to the professional person are believed to unduly outweigh the benefits to the client or the general public, then demands are made for some sort of regulation. We have argued that the expectations we each hold about other people are, at bottom, expressions of our belief that behaviour is both predictable and trustworthy. There is a trusting relationship between the general public and those who are members of a defined professional group. In short, privilege and the status which goes with it has to be paid for in demonstrating trustworthy action or behaviour associated with the common good. That is to say, professional people are expected to make a positive contribution to the well-being of the society in which they operate.

The mutual expectations implicit in the two sets of relationships described, the one relating to the individual patient and the other to society as a whole are matched by yet a third set relating to the expectations which exist between professional colleagues. A balancing of these three sets of expectations with the professional person's own personal feelings and beliefs is difficult to achieve and conflicts sometimes arise. This potential source of conflict is not new to the medical profession; the abortion debate which still raises considerable disagreement both within and outside the medical profession is an example. Nevertheless, balancing the perceived 'needs' of the patient, one's colleagues, society, and personal conscience, all at the same time, remains a formidable undertaking.

In a highly individualistic culture such as ours and a particularly individualistic profession such as medicine, the relationship between doctor and patient is usually a highly confidential one. In normal circumstances the patient has a need which can be defined in medical terms and the doctor is expected to be able to meet that need from within his or her professional competence. But some needs go beyond a purely medical interpretation; they may be described as social needs. We would argue that the provision of AID to assist a healthy woman to achieve a pregnancy is an example of meeting a social need rather than a medical one; the 'treating' of a fit wife for an incapacity in her husband is hard to justify in purely medical terms. When faced with the problem of male infertility, AID is being used not to treat infertility but merely to circumvent it; AID may cure a man's childlessness but it does not cure his infertility. What is being met is a *social* need rather than a purely medical need. If AID was not such a comparatively easy way of dealing with childlessness, perhaps medical research into male infertility might be further advanced.

If the notion of social need is accepted then the doctor's role is enlarged; professions like other occupational groups can be imperial in their outlook. The strictly medical justification for the doctor's involvement in AID provision is weak. Once the initial gynaecological examination to determine the causes of infertility has confirmed that a woman is potentially fertile, the timing of ovulation is the main medical procedure required, and this can be done by those who are not medically qualified. Nor is the selection of a donor a strictly medical procedure. The main reason for the doctor's involvement in AID provision appears to be more social than medical. The doctor acts as an 'honest broker' mediating between the couple and a suitable donor, and tends to legitimise a procedure which requires a high level of confidence by the couple seeking a pregnancy and by the donor who is providing the semen. Of course, some forms of artificial reproduction are more complex than artificial insemination. External human fertilisation, for example, can only be performed by highly skilled professional people but AID, the most common procedure, is essentially simple to undertake and can be provided by almost anyone and even on a 'do-it-yourself' basis. To make AID a medical matter, the procedures are usually presented partly as an extension of a gynaecological examination, partly as an application of technology such as sperm banks and the freezing of semen in liquid nitrogen, and partly by the use of technical and nursing assistance. In some respects the presentation of such services as medical services within a medical setting is a charade; but there is a need, on the part of individuals and of society, for services to be provided in confidence by a member of a profession which can be trusted. Clearly, a compromise based upon a carefully constructed and agreed framework is required which will protect the interests of the childless patient, the doctor and society as a whole.

Up to this point we have discussed the professional position of the clinician providing an artificial reproduction service within the wider community for the individual patient. There is another area of potential conflict but this time within the profession itself. This relates to the selection of semen donors. It is frequently asserted that most donors are medical students. There is some evidence to support this statement although it is known that donors who are not members of the medical profession are also used. The use of medical students means that these young men are exposed to subtle forms of pressure. As new entrants to a profession for which they have a high regard and from which they have high expectations, requests from senior members of that profession may exert an undue pressure. What motivates a medical student to agree to donate semen? To be sure, there is usually a sum of money provided for each ejaculate, but the amount is generally insufficiently large to act as the sole means of motivation. Another commonly mentioned source of donors is that of fertile husbands whose wives have been successfully treated for infertility. Here a feeling of gratitude and obligation to the doctor may also exert an undue pressure to comply. But the provision of money, and other less obvious forms of persuasion, raise the question of donor selection and donor accountability. The eugenic implications of donor selection, and the psychological and social implications of partaking in the responsibility for the creation of a new human being, are matters which though poorly understood at the present time are of considerable importance to an understanding of the social implications of artificial reproduction. An anonymous donor must forego any knowledge of or interest in his genetic offspring; rights, duties and obligations must be denied. The donor's kin are similarly affected and are denied knowledge of 'blood' relatives; a donor's mother may have unknown grandchildren or a donor's children may have unknown half-siblings. The rights, duties and obligations of family life are deeply ingrained and not easily denied. It is essential that potential donors should be able to explore the implications of their decision with a skilled counsellor. In addition, how donors are selected, how often they are used and whether or not payment for semen should be made are issues that attract a wider consideration than by those who are providing or using artificial reproduction procedures.

ARTIFICIAL REPRODUCTION AND RESEARCH

The doctor providing a service for childless couples may

need to co-operate with professionals working in other disciplines if a full and complete service is to be provided. Each professional person contributes a particular expertise which, combined with his or her own motivations, may be a source of potential inter-professional conflict. Whereas the doctor may wish to provide a service for an individual patient or couple, the research scientist may have other imperatives uppermost in mind. Research scientists are driven by strong desires for progress in advancing know-ledge in their own particular discipline and to do this they may well want to use procedures or even live human tissue in ways that raise concern for the clinician and other members of society. A recent report reminds us:

> There exists a strong technological impetus, a desire to see how things work, to analyse and experiment. Human beings display an innate curiosity, the source of so much that is good in human life. However, this impetus can turn into a categorical imperative. If something is possible, it must and will be done. Curiosity becomes limitless, a compulsion to do whatever is possible regardless of the consequences. However, technological power does not guarantee truly human solutions to the deepest human problems. Cleverness is not the same thing as wisdom. The question where to place limits, where to draw the line, remains as important as ever.

This 'technological imperative' may be closely related to a 'consumer imperative'. We all have a strong drive to satisfy our felt wants. Consumerism is there to stimulate our wants. If we are not careful, we cease to question the values underlying our wants. We equate our wants with our needs and our rights. And technology is there to give us what we want. But we ought to ask questions of value. We ought to ask what constitutes 'the good life', what it is that fulfils people's deep humanity. (Free Church Federal Council and The British Council of Churches, 1981)

It is clear that not only those providing a direct service, but also those who undertake the research which underlies it, have a responsibility to society as a whole. The need for the researcher to demonstrate the positive benefits to society of the research undertaken is particularly important. The uncertainty and even suspicion that has been roused by artificial reproduction has been partly due to a lack of accurate information about the research undertaken. It is possible that such uncertainty is unwarranted but it is not surprising that it has been generated. Incomplete snippets of information have been publicised under newspaper head-lines such as 'Waiting for a Sperm-bank Genius', 'Clones in a Synthetic Womb', 'Embryo Tissue for Use in Transplants Urged'; the correspondence and leader columns in national newspapers and medical journals all indicate that there has been a breakdown in confidence that particular research activity was being undertaken for the benefit of society as a whole.

The build-up of social concern about the implications of artificial reproduction research and practice resulted in the appointment of a Committee of Inquiry in 1982. Had the development of these new techniques not been surrounded by such an element of secrecy perhaps the need for an official inquiry of this sort would have been unnecessary. It was not just concern about the procedures of artificial reproduction that led to the need for an inquiry, but also a lack of infor-mation, fears that insufficient control was being exercised, and uncertainty about the social implications of the proce-dures. The committee members were chosen so as to repre-sent a wide cross-section of the general public, thus avoiding any suggestion that only a narrowly professional point of view was being expressed. Evidence was invited from a wide range of individuals, organisations and public bodies. Similarly the Inquiry's terms of reference were drawn up to demonstrate the broad social concern that artificial forms of reproduction had aroused.

Artificial Reproduction and Society

To consider recent and potential developments in medicine and science related to human fertilisation and embryology; to consider what policies and safeguards should be applied, including consideration of the social, ethical and legal implications of their developments and to make recommendations.

The establishment of the Committee of Inquiry is a clear example of the perceived relationship between the needs of a more narrowly defined professional group and those of the wider society of which the professional group is a part. This serves as a reminder that members of a profession or scientists engaged in research are also members of the general public and while professionalism may require conformity with a particular code of practice the professional is not absolved from the more general social, legal and ethical responsibilities which membership of the wider society demands.

9

Conclusions and Recommendations

In this chapter we wish to summarise our findings based on researches into AID and to generalise them to artificial reproduction of all kinds. In so doing we make some specific recommendations which concentrate on the social implications of artificial reproduction. These recommendations are discussed more fully in the previous chapters of this book and this chapter should not be read in isolation. Artificial reproduction, other than in the case where a husband's sperm fertilises his wife's ovum, is, from a social standpoint, disruptive. This will remain so even if the problems associated with the secrecy surrounding the provision of artificial reproduction services were to be fully resolved. However, some forms of artificial reproduction such as AID and external fertilisation are being provided on an increasing scale and they are unlikely to be suppressed. Clearly, our society must find ways of accommodating itself to these new forms of reproduction. But this does not mean that some sort of control is out of the question. Indeed our main conclusion is the desirability of developing guidelines for these practices, some of which may need to be the subject of legislation, or encapsulated in a professional code of practice. There must

166

be a formal recognition of the moral responsibilities required of the individuals providing and using the services available, and of the social rights of the children created in this way.

In presenting the results of our researches and our reflections upon them we argue that artificial reproduction, whilst meeting the needs of some, but still relatively few, members of society has also to be considered in terms of its effects, or likely effects on society at large. In order to provide social control of some sort, a framework must be constructed within which the practice of artificial reproduction can effectively be supported. In constructing such a framework we do not all have to agree to the particular form it takes, nor do we have to see it as permanent. The abortion debate is an example where uncertainty still exists in many minds but the meeting of the needs of individuals within a socially acceptable framework of regulation has been more or less achieved. The present arrangement for abortion may not satisfy everyone but it does permit a need to be met in ways that are preferable to the alternatives that obtained previously.

RECOMMENDATION I.   A formal structure should be established in which artificial reproduction can be practised in a way which is not deleterious to society.

TERMINOLOGY

One of the problems affecting an accurate identification of the social implications of various forms of artificial reproduction has been the lack of an adequate terminology. Describing procedures without reference to the roles certain individuals are playing when using those procedures has led to a confusion of definitions with the result that what is meant by one person is often not being fully understood by another. More procedures are possible than those described in this book but those defined in Table 2.2 are the most likely, and refer to Recommendations III–VI. Justification for this nomenclature is given in Chapter 2.

RECOMMENDATION II.   To avoid confusion in terminology the following nomenclature should be adopted in discussions of artificial reproduction:

A.   *Roles and Relationships of Parents Using Artificial Reproduction*

 a) *Mother*
   i) Genetic mother ⎫
   ii) Carrying mother ⎬ Complete mother
   iii) Nurturing mother ⎭

 b) *Combined roles of mother*
   i) Genetic/carrying mother
   ii) Carrying/nurturing mother
   iii) Genetic/nurturing mother

 c) *Father*
   i) Genetic father ⎫
   ii) Nurturing father ⎬ Complete father

B.   *Artificial Reproduction Procedures*

 a) *Artificial insemination*
   i) By husband (AIH)
   ii) By donor (AID)

 b) *External human fertilisation* (EHF)
   i) Embryo replacement
   ii) Embryo transfer

MARRIAGE

Because the family is a basic organisation in society and provides the social milieu where children are brought up from infancy to accept the fundamental values of society, it is appropriate that children born by means of artificial reproduction should be born within the socially sanctioned bond of marriage. The fact that some naturally conceived children are born outside of marriage and that normal parenthood can be responsible for putting children at risk is no justification for deliberately setting up conditions of risk for them. If marriage is eliminated there is no framework within which social control can be exercised over the responsibilities society has for the child. It follows that to provide artificial reproduction for women outside marriage is deliberately to bring about the birth of children who are in an abnormal situation, creating problems of child-care and identity, and introducing more complications into an already confused situation.

RECOMMENDATION III. Artificial reproduction should only take place where a couple responsible for nurturing the child are married.

COMPLETE PARENTHOOD

Given that artificial reproduction is acceptable, it is desirable that at least one parent should fulfil all the reproductive roles normally undertaken by the parent of that sex. To have two incomplete parents introduces unnecessary confusion into the family. Not to restrict artificial reproduction to a couple where at least one is a complete parent is to lose control of the reproductive process and to proliferate familial roles to a point where a child will be uncertain of his or her relationships.

RECOMMENDATION IV.   At least one of the married partners should be a complete parent, i.e. a complete mother incorporating the genetic, carrying and nurturing

169

functions of motherhood, or a complete father, incorporating the genetic and nurturing functions of fatherhood.

COMMITMENT TO ACCEPT THE CHILD

The birth of a handicapped child creates great stress for any couple, but the birth of a handicapped child following artificial reproduction holds the potential for even greater stress. Recipient couples require counselling to explore their potential reaction to this unlikely, but nevertheless possible outcome. In the normal situation a married couple have a continuing responsibility for their child irrespective of its condition at birth and subsequently. The same should be no less so for couples who have elected to have a child by any other means. Some legal provision will be needed to meet this requirement.

RECOMMENDATION V. The couple should be required to make a legal declaration in advance of artificial reproduction that they will both accept the resulting child as their legitimate offspring.

SURROGATE MOTHERHOOD AND WOMB-LEASING

Those forms of artificial reproduction which require the services of a woman who gives up the child at birth to be nurtured by someone else give rise to particular concern. Unlike adoption, where a nurturing mother takes over the care of an already existing child produced by a genetic-carrying mother, surrogate motherhood or womb-leasing techniques are purposefully used to create a child in a deliberately planned way which introduces unnecessary social confusion for the child. It is known that the developing relationship between mother and child during pregnancy is of considerable importance for the future well-being of the child. To break this bond is traumatic to the mother also.

RECOMMENDATION VI. The carrying of a child, following artificial reproduction, should be disallowed where the carrying mother is not also to be responsible for the nurture of the child.

ARTIFICIAL INSEMINATION BY DONOR

Where a married woman is capable of conceiving, carrying and nurturing a child but her husband is infertile, subfertile or is a carrier of an hereditary disease, semen donated by a male donor and introduced by artificial insemination techniques is often used to effect a pregnancy. This is the most common form of artificial reproduction and presents an opportunity to examine and predict the social implications of other more complex forms of artificial reproduction.

RECOMMENDATION VII. AID should be permitted in accordance with appropriate safeguards.

COMBINING AIH AND AID

To combine husband's semen with that of a donor is to introduce social confusion and self-deception. To mix the semen not only reduces the chance of fertilisation but encourages the husband and wife to deny the role of the donor. This practice also compounds the confusion of the child with regard to its genetic inheritance.

RECOMMENDATION VIII. The practice of combining husband's and donor semen should be discouraged.

OVUM DONATION

In those cases where a wife is capable of carrying and nurturing a child but does not ovulate, or may pass on hereditable disease, it is possible for her to receive a donated ovum from another woman. In such cases the procedure would be for the husband's sperm to fertilise the donated ovum, the resulting embryo being transferred to his wife's uterus. This process is the female equivalent of AID and is subject to the same social considerations.

RECOMMENDATION IX. The donation of an ovum by a female donor, which is then transferred to a wife's uterus following fertilisation using the sperm of her husband, should be permitted under similar safeguards to those recommended for AID.

PROVISION OF SERVICES

One of the most important features of a framework designed to support social control of artificial reproduction is that associated with the provision of the service. It has been shown that artificial reproduction is a potential threat to basic values in society. It is therefore imperative that such social control as is necessary should be exercised by a responsible body. The medical profession appears to be the most suitable body. Being a profession known to have high ethical standards, it is already providing a service, and artificial reproduction is a natural extension of the diagnosis and treatment of infertility. Because there is at present some divergence in the kinds of services provided, it is necessary that a code of practice having specific relevance to artificial reproduction should be implemented, and should include some form of registration of those engaged in this work.

172

RECOMMENDATION X. Those practising artificial reproduction techniques should be registered and should be expected to conform to a specific code of practice. This indicates the need to retain such services within the medical profession.

In adoption, couples seeking a child are carefully counselled in order to allow them to discuss with a skilled and supportive person if this means of dealing with childlessness is the most appropriate for them and for the child. While adoption is not an exact parallel to AID, the desirability for counselling is no less necessary. The decision by a married couple to accept artificial reproduction techniques to alleviate childlessness must be securely based on an adequate exploration of the social implications. A skilled counsellor is necessary to guide and direct this exploration. Using the adoption experience as a starting point the counselling service should also cover the present and future difficulties which may be associated with secrecy, both with respect to kinsfolk and the child itself. Ways should be found of providing help to parents concerning the most appropriate time and manner of informing the child of its different genetic origin. The 'scripts' provided by adoption agencies may be used as a model.

RECOMMENDATION XI. Before accepting artificial reproduction techniques a married couple should receive skilled counselling. This counselling should include the following topics:

(a) alternative ways of coping with childlessness;
(b) encouragement to include close relatives in the decision;
(c) encouragement and support in informing the child of its genetic origins in a non-damaging way.

173

Artificial Reproduction

INDEPENDENCE OF THE COUNSELLING SERVICE

Counselling for couples prior to artificial reproduction should be provided by skilled people who are professionally independent of those providing the medical service. This counselling whilst mandatory should not be an assessment, neither should it curtail the right of the doctor to select couples. The skills of such counsellors will lie in a different direction from that of the doctor and would relieve the medical team from a time-consuming task. Moreover, it is desirable that such a service should be continuously available. Experience shows that AID couples value an opportunity to discuss issues relating to their situation at various times in the life of the child, and they do not feel necessarily that the medical service is appropriate for this.

RECOMMENDATION XII. Counselling the couple should be undertaken by trained people who are independent of the medical service to be provided. Such a counselling service should be continuously available to those who have had artificial reproduction in the past.

DONOR COUNSELLING AND ASSESSMENT

Before donating semen or ova, donors should be counselled and assessed as to their suitability by someone skilled in the task and who is not directly connected with the team providing the service. This is necessary to avoid pressure being brought to bear on a person to become a donor which they may later regret. It is only after counselling and independent assessment that a medical examination should take place. The donor should declare in writing his or her willingness to help and where the donor is married the spouse should also assent. The donor should be told what information will be made available to any subsequent children, and he should have no legal rights or obligations in relation to such children.

174

The donor should be assured that his or her identity will not be revealed. Donors should receive no payment apart from expenses for their help.

RECOMMENDATION XIII. Donors must be counselled and assessed before providing semen or ova. Where the donor is married the spouse, in addition to the donor, should give written consent. Donors should not be paid.

FREQUENCY OF USE OF DONORS

There is a frequently expressed fear both by parents experiencing artificial reproduction ānd by members of the general public that half-siblings conceived from the same donor and unaware of their genetic relationship may meet as adults and marry. This possibility may be unlikely but the social and psychological fears expressed are nevertheless present and have an important effect on confidence in artificial reproduction services. The statistical probability of such an occurrence can be calculated and this may give reassurance. For this to be done an upper limit to the number of offspring permitted from each donor must be defined, and for this calculation to be meaningful the upper limit must be observed. Eugenic considerations also make it desirable for the number of conceptions from each donor to be limited.

RECOMMENDATION XIV. Conceptions arising from one donor should be limited to a number agreed within the medical code of practice.

MATCHING OF DONOR AND RECIPIENT

At present it is the practice for AID practitioners to attempt to match, as far as is possible, the physical characteristics of the donor to those of the husband. It could be presumed that if couples were encouraged to be less secretive about AID this matching would be unnecessary. However, experience with adopted children has shown that a physical likeness to the adoptive parents helps the child to identify with the parents, and the parents with the child. Adopted children, whilst not wishing their adoptive status to be secret, desire it to be taken for granted so that it can then be forgotten; an incompatible physical appearance would hinder this. It is therefore desirable to attempt to match donors and recipient parents.

RECOMMENDATION XV. Donors of semen or ova should be matched for physical characteristics, as far as is practicable, with the recipient parent.

LEGITIMACY OF THE CHILD

In the light of previous recommendations it is inconsistent that the parents of a child resulting from artificial reproduction should have to engage in a subterfuge when registering the birth of the child. Provision should be made to enable the child to be registered as the legitimate offspring of the nurturing parents.

RECOMMENDATION XVI. A child born as the result of artificial reproduction techniques should in law be recognised as the legitimate child of the nurturing parents.

AVAILABILITY TO THE CHILD OF THE DONOR'S MEDICAL AND SOCIAL BACKGROUND

In the deliberate creation of a child using donated semen or a donated ovum there is no justification for the artificially-reproduced child to be deprived of all knowledge of its

176

genetic history. This does not require the identity of the donor to be revealed. A medical history and a pen-profile of the donor should be made available to the parents, and at an appropriate time to the child.

RECOMMENDATION XVII. **Every child born by artificial reproduction should have its genetic history recorded and this should be made available to the child's parents and to the child at an appropriate time. The genetic and social background of the donor should not include the means of identifying the donor.**

REGISTRATION OF CENTRES

In order to ensure social control of artificial reproduction services it is necessary that they are provided at identified and registered centres. Where such services include facilities for freeze-storage of human tissue the centre should be licensed for the purpose and be periodically inspected. Charges to couples seeking artificial reproduction should be in conformity with the medical code of practice.

RECOMMENDATION XVIII. **Centres where artificial reproduction techniques are provided should be registered. Facilities for the freeze-storage of semen, ova or embryos should be licensed.**

EXPERIMENTATION ON HUMAN EMBRYOS

It was noted in Chapter 1 that the relatively simple procedures heralded by AIH have led to scientific developments hardly imaginable at the time. The major steps in this development relate to the deep-freezing and storing of semen and embryos and to the ability to fertilise an ovum outside the human body. We have argued more than once in this book that the delivery of a husband's semen or the replacement of a woman's ovum after external fertilisation using her husband's semen is of little social consequence. Where donors are involved or where the couple are not married, social questions begin to arise. These questions apply mainly to the rights of the child yet to be born. The rights so far discussed relate to the child's identified position within the family and within society generally; a position which should be clear and unambiguous and which should not be based on deception. But the child has rights concerning its own physical constitution which are no less important than its social development.

External fertilisation has led to scientific opportunities which in one way or another require manipulation of the human embryo. The ethical problems posed by the possibility of experimentation upon human embryos are very complex and a detailed study is outside the remit of this book. The examination of embryos at a very early stage in their development can be beneficial in advancing knowledge and thereby providing help to improve the quality of life for others. Yet this is an assumption which must be examined very carefully, for the material acting as the subject of the experimentation is a human being at the beginning of its individual development. The issues surrounding such experimentation may involve questions of an ethical or philosophical nature, but social questions arise in relation to the control society wishes to exercise over those undertaking such research on its behalf. Examples of possible experimentation include use of embryos to test new drugs or other substances for eventual use by humans; genetic manipulation; the development of the embryo outside the uterus; cross-species fertilisation; and the development of tissue and organs for transplantation. These experimental possibilities are being discussed at the present time and the pace of scientific advance suggests that they will be more than possibilities within the foreseeable future. It is clear that much of this experimentation will be for the good of individuals in need, but uncertainty remains because of a lack of control over experimentation, and the possible misuse that could so easily take place. With hindsight, those living in the next century may regret developments that permit such possibilities as sex pre-selection of children, or posthumous use of donated semen, ova or embryos. The positive advantage of removing inherited disease or congenital deformity have to be seen in the light of the possible disadvantages of genetic manipulation.

Experimentation on human embryos is quite different from experimentation on dead human tissue or living animal tissue. The value placed upon human life is so great that experimentation on human embryos should only be undertaken if it can be shown to be in the public interest. Moreover, without some control the strong desire of scientists for new knowledge may lead them to undertake work which is progressively removed from human and social interests. Professional codes of practice are useful but insufficient in this sphere unless there is public enforcement.

RECOMMENDATION XIX. Experimentation on human embryos should only be undertaken by those who are registered and who have agreed to work within a defined and agreed code of practice; their work should be subject to official inspection.

NEED FOR CONTINUOUS SURVEILLANCE

In this book we have tried to show that the relationship of the child to others within the family is linked to the relationships also found in the wider society. It is argued that by finding ways to remove the secrecy and deception usually associated with artificial reproduction where donor gametes

179

are used, much of the difficulty surrounding its use would be removed. This is apparent when the situation is viewed from the social perspective; but for the husband and wife involved, the decision to be open about their action cannot be an easy one, especially when there is uncertainty about how others will react to the news. The conflict between personal need stemming from the desire to be seen as no different from other people, and the requirements of an acceptable social structure is hard to resolve. Some will, no doubt, argue that such conflict is irreconcilable and that secrecy is the only means of avoiding it. This may have been so before public awareness of some of the issues associated with AID and external human fertilisation became apparent. Now that the Pandora's Box has been opened there is no going back; the best that can be hoped for is a framework within which both personal need and social confidence can be supported. Doubtless, some will find the framework restricting especially in relation to the recommendations on marriage (Recommendation III) and the wish to contain the service within the medical profession (Recommendation X). Equally, there will be those who believe the framework is too lax. One thing which is certain is the lack of effective social control at the present time. Sperm and embryo banks, the commercialisation of artificial insemination, the use of 'surrogate' carrying mothers and the provision of external human fertilisation services are all currently unregulated. Such services are directed towards those who are particularly vulnerable in their desire to have a child almost at any price. The depth of misery and the feelings of hopelessness experienced by couples who want a child but, for one reason or another, are denied one, are very evident. The need for regulation is not in order to prevent these couples from receiving the help they need, but to ensure that such help is provided in a way that is compatible with the moral and ethical values of our time and avoids the dangers which accompany secrecy and the lack of regulation.

The need of individual couples has also to be seen in terms of the experimentation that accompanies the services being provided for them. Concern about future developments such as the fractionating of semen to provide numbers of possible conceptions from a single ejaculate, the freezing of embryos for longer and longer periods of time, the ability to grow an embryo for longer periods outside the human body, and the advances being made in replicating cells are not based on irrational fears but on an awareness of what has already taken place. It is therefore of profound importance that social control of artificial reproduction is implemented through the process of open debate and agreed regulation. It is also important that the review body established should have a continuing function in regulating this important area of human behaviour.

RECOMMENDATION XX. A socially representative review body should be established with responsibility for a continuous surveillance of research and practice in the sphere of artificial reproduction.

CONCLUDING NOTE

This social investigation, whilst focusing on artificial reproduction, has raised a number of issues that go far beyond the subject itself. By discussing the pros and cons of these practices we have found ourselves asking questions of larger import. Thus the nature of the family in modern society has been a topic of discussion. The relevance of wider kinsfolk and the part played by the institution of marriage have also been considered. All this leads us to think afresh about family life, not just in relation to artificial reproduction, but generally in modern urban industrial society.

To be sure time brings changes but only because change is

endemic to large-scale societies, and changes in one sphere will have an influence on other spheres of human life and activity. In considering artificial reproduction the authors of this book have become aware that further studies are needed within psychological and sociological frames of reference into family life and the normative order governing it. Artificial reproduction is not, therefore, an independent and specific phenomenon; it raises a host of questions by implication. Whilst this is not the place to raise them, it should be clear that further work needs to be done if we are to understand more clearly the domestic world in which we live so much of our lives. Conjugal roles, the effects of technology on sexual relationships, the roles of women, family structure and the relationships of parents and children are among the topics to be considered, and to be thought about both from the point of view of the family group and also the wider society. The separation of reproduction from sexual intercourse has implications for the basic structure of human society. It is not unreasonable to believe that this study has helped to focus more closely on these and related topics.

Bibliography

Archbishop of Canterbury (1948), Report of a Commission Appointed by His Grace the Archbishop of Canterbury, *Artificial Human Insemination* (London: SPCK).

Bok, S. (1978), *Lying: Moral Choice in Public and Private Life* (Hassocks, Sussex: Harvester Press).

Brandon, J. (1979), 'Telling the AID child', *Adoption and Fostering*, vol. 95, no. 1, pp. 13–14.

Czyba, J. C., and Chevret, M. (1979), 'Psychological reactions of couples to artificial insemination with donor sperm', *International Journal of Fertility*, vol. 24, no. 4, pp. 240–5.

David, G., and Price, W. S. (1980), *Human Artificial Insemination and Semen Preservation*. New York: Plenum Press.

Dunstan, G. R. (1973), 'Moral and social issues arising from AID', in G. E. W. Wolstenholme, and D. W. Fitzsimons (eds), *Law and Ethics of AID and Embryo Transfer*, CIBA Foundation Symposium 17 (New Series) (Amsterdam: Associated Scientific Publishers).

Dunstan, G. R. (1975), 'Ethical aspects of donor insemination', *Journal of Medical Ethics*, vol. 1, pp. 42–4.

Edwards, R. G. (1980), *Conception in the Human Female* (London and New York: Academic Press).

Edwards, R. G., and Purdy, J. M. (eds) (1982), *Human Conception in Vitro*, Proceedings of the first Bourn Hall meeting (London: Academic Press).

Feversham, Lord (1960), Report of the Departmental Committee on Human Artificial Insemination (Feversham Report), Cmnd 1105 (London: HMSO).

Free Church Federal Council and The British Council of Churches (1981), *Choices in Childlessness*, Report of a Working Party.

Goffman, E. (1963), *Stigma* (Englewood Cliffs, N.J.: Prentice-Hall).

Grobstein, C. (1981), *From Chance to Purpose: An Appraisal of*

External Human Fertilisation. (Reading, Mass.: Addison-Wesley).

Hafez, E. S. E., and Semm, K. (eds) (1982), *In Vitro Fertilization and Embryo Transfer* (Lancaster: MTP Press).

Hanmer, J. (1981), 'Sex predetermination, artificial insemination and the maintenance of male dominated culture', In H. Roberts (ed.), *Women, Health and Reproduction* (London: Routledge & Kegan Paul).

Holmes, H. B., Hoskins, B. B., and Gross, M. (eds) (1981), *The Custom-Made Child? Women-Centered Perspectives* (Clifton, N.J.: Humana Press).

Howard, T., and Rifkin, J. (1977), *Who Should Play God?* (New York: Delacorte Press).

Jackson, M., and Richardson, D. (1977), 'The use of fresh and frozen semen and human artificial insemination', *Journal of Biosocial Science*, vol. 9, no. 2, pp. 251–62.

Katz, I. (1981), *Stigma. A Social Psychological Analysis* (Hillsdale, N. J.: Lawrence Erlbaum Associates).

Keane, N. P., with Breo, D. L. (1981), *The Surrogate Mother* (New York: Everest House).

Lambert, L., and Streather, J. (1980), *Children in changing families* (London: Macmillan).

Law Commission (1979), *Family Law: Illegitimacy*, Working Paper no. 74 (London: HMSO).

Law Commission (1982), *Family Law: Illegitimacy* (London: HMSO).

Ledward, R. S., Symonds, E. M., and Eynon, S. (1982), 'Social and environmental factors as criteria for success in AID', *Journal of Biosocial Science*, vol. 14, no. 3, pp. 263–75.

McWhinnie, A. M. (1967), *Adopted Children. How they grow up* (London: Routledge & Kegan Paul).

Oxtoby, M. (ed.) (1981), *Medical Practice in Adoption and Fostering* (London: British Agencies for Adoption and Fostering).

Payne, J. (1978), 'Talking about children: an examination of accounts about reproduction and family life', *Journal of Biosocial Science*, vol. 10, no. 4, pp. 367–74.

Peel, Sir John (1973), Report of the Panel on Human Artificial Insemination (Peel Report), *British Medical Journal*, vol. 2, Supplementary Appendix V, 3.

Bibliography

Peyser, M. R., Ayalon, D., Harell, A., Toaft, R., and Cordova, T. (1973), 'Stress-induced delay in ovulation', *Obstetrics and Gynecology*, N.Y., vol. 42, p. 667.

Royal College of Obstetricians and Gynaecologists (1979), *Artificial Insemination* (London: RCOG).

Smith, L. A. (1980), 'Artificial insemination: disclosure issues', *Columbia Human Rights Law Review*, vol. 11 (Spring/Summer), pp. 87–101.

Wolkind, S. (ed.) (1979), *Medical Aspects of Adoption and Foster-Care* (London: Heinemann Medical).

Walters, W., and Singer, P. (eds) (1983), *Test Tube Babies: A Guide to Moral Questions, Present Techniques and Future Possibilities* (Melbourne: OUP).

Index